It's Dogwood Time in Tyler County!

What is **Spring** in Tyler County, Texas? A glorious abundance of beautiful blooming trees and flowers against a dark background of towering pines. With a community purpose, spirit and vigor, Tyler Countians have paid tribute to the glories of Spring and the lovely dogwood trees since 1940 by creating each year a growing and more beautiful Dogwood Festival.

In an invitation address written in 1941 by Judge James E. Wheat, founding father of The Dogwood Festival, he bestowed honor upon the Dogwood tree and expressed the spirit of Spring in Tyler County. He wrote:

"For all of these years, the faithful dogwood has proclaimed the passing of Winter and the coming of Spring. Each succeeding generation has enjoyed it's beauty and looked upon it as a good one. From its opening, the farmer has decided when to plant his crops, the sportsman when to go fishing and the politician when to announce for office."

"It has seen soldiers march off to war. It has seen them return—sometimes triumphant and sometimes defeated. It has seen the giants of the forest—the oak, the pine, and the magnolia removed. It has survived the good years and the bad years, the successes and the failures of the people, their joys and their sorrows. Through it all, it has been among the first to burst forth in it's whiteness in the Spring."

The Dogwood Festival has grown from a small one day occasion to three weekends of endless activities and pleasures against the background of East Texas Springtime.

DOGWOOD BEGINNINGS:

The idea to present the festival actually originated in 1938. That year citizens of Tyler County met near the Tyler-Polk County line with officials from the State Highway Department to make arrangements for the construction of Highway 190. At the time, Judge Wheat made a suggestion that the beauty of East Texas in Spring-time should be publicly celebrated in some way that visitors from all over the state might be able to enjoy the sight of the dogwood trees in bloom.

The suggestion was brought to life, and the first festival was held on April 6, 1940, under the sponsorship of the Tyler County Chamber of Commerce.

-By: Vicki Gassiott

Welcome to the 73RD Annual Tyler County DOGWOOD FESTIVAL

From the Woods

The words of the late James E. Wheat,
Our Dogwood Festival's founding father, still ring true today.

An Appreciation
and
Acknowledgement

The loyal cooperation, hard work and enthusiastic assistance of the people of Tyler County in making the Dogwood Festival one of the most beautiful and unique shows of East Texas, is sincerely appreciated.

This demonstrates what can be done in small towns when there is a spirit of good will, broadminded appreciation of the cultural and esthetic values of life and community spirit that enables people to work together for the common good.

We also wish to gratefully acknowledge the spirit of helpfulness and cooperation demonstrated by our neighboring counties. We have tried to produce outstanding performances and have availed ourselves of the best talent in Southeast Texas.

To all our friends everywhere, we say thank you.

Sincerely,
The Tyler County Dogwood Festival
Officers & Directors

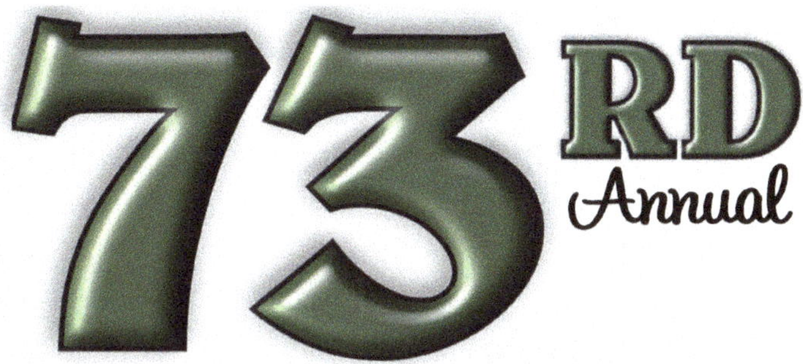

Tyler County DOGWOOD FESTIVAL
From the Woods

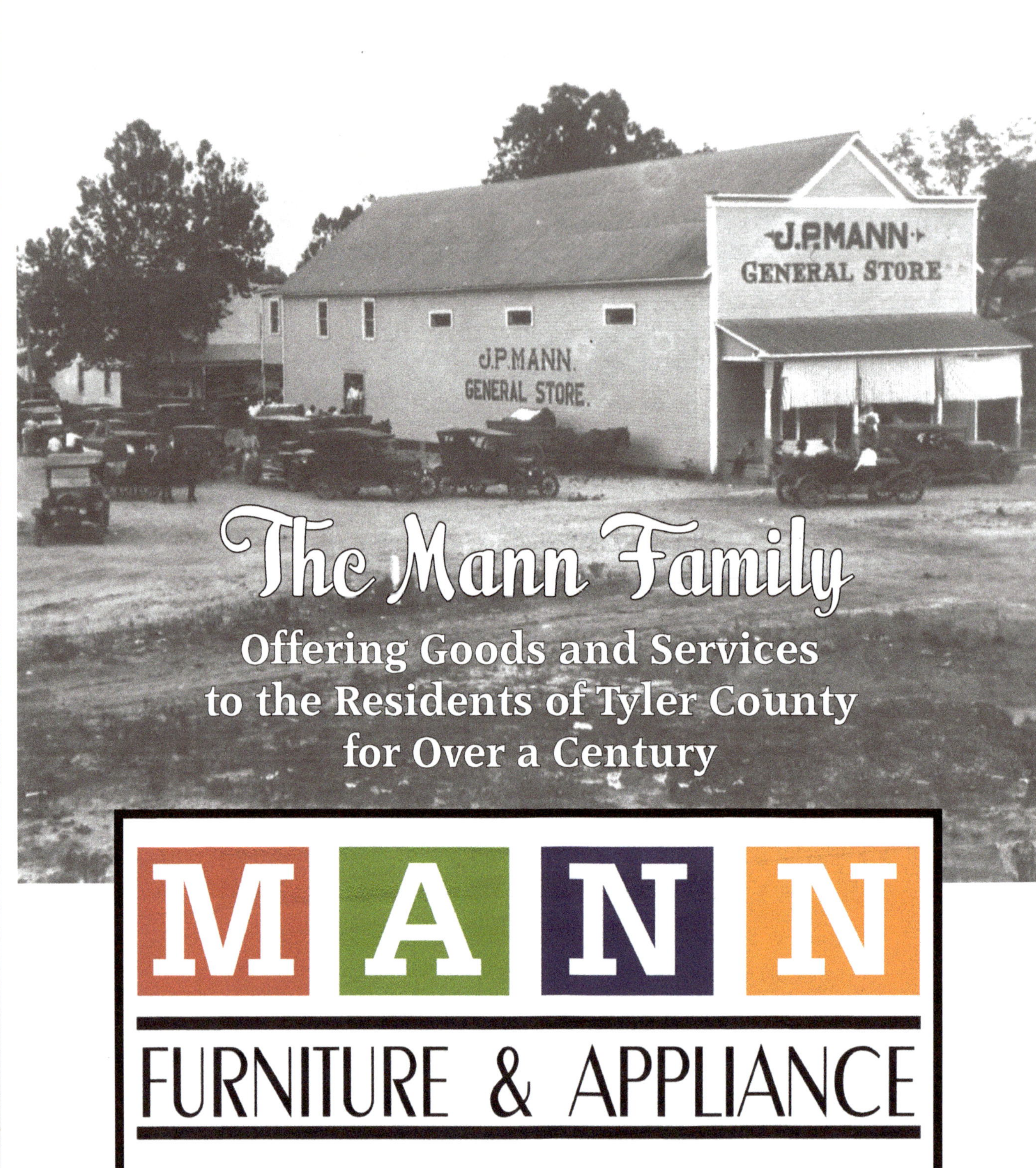

J.P. MANN
GENERAL STORE
J.P. MANN.
GENERAL STORE.
The Mann Family
Offering Goods and Services
to the Residents of Tyler County
for Over a Century
MANN
FURNITURE & APPLIANCE
1123 SOUTH MAGNOLIA
Woodville, TX   409-283-8286

Tyler County Dogwood Festival
SCHEDULE OF ACTIVITIES

Festival of the Arts Weekend

Friday • March 18

6:30pm-9:30pm...A Night at the Tyler County Courthouse
First Annual Gala sponsored by the Tyler County Historical Commission
Tickets available at Heritage Village Gift Shop • www.tylercountyhc.org

Saturday • March 19

9:00am-5:00pm...Heritage Village • 409-283-2272
Village Tour, Dogwood Festival Exhibit, Spinning and Weaving, Quilt Show,
Buggies & Wagon Exhibit, Chair Caning/Cattail Rush Weaving, Railroad Museum,
Wild Thymes Herb Farm, Bluegrass, Gospel & Folk Music, James and Priscilla Hale
Located 1 mile west of Woodville on Hwy 190
Free Admission

10:00am-4:00pm. Tyler County on Tour • Self-Paced Driving Tour • Maps available at Heritage Village
Adults: $10.00 ea • Ages 6-17: $4.00 ea

Sunday • March 20

11:00am-2:00pm. Heritage Village Dinner on the Grounds with Live Entertainment
Adults: $10.00 ea • Children under 12: $5.00 ea

2:00pm-4:00pm...Royal Tea • Meet the 2016 Dogwood Royalty! • Village Street Bed & Breakfast
Each visitor receives a crown, sash and picture.
$20.00 each

Western Weekend

Friday • March 25

7:00pm.................Rodeo
Tyler County 4H/FFA Rodeo Grounds
Sponsored by the Woodville Lions Club

Saturday • March 26

8:00am.................Sweetheart Horsemanship Competition
Tyler County 4H/FFA Rodeo Grounds

9:00am-5:00pm...Arts & Crafts Show
Downtown Woodville
Sponsored by
Tyler County Texas Business Women

2:00pm.................Western Trailride Parade
Downtown Woodville

4:30pm.................Rodeo • Pre-Rodeo begins at 4:00pm
Tyler County 4H/FFA Rodeo Grounds
Sponsored by the Woodville Lions Club

8:00pm-12:00am. Western Dance
4H Building
Tyler County 4H/FFA Rodeo Grounds

Queen's Weekend

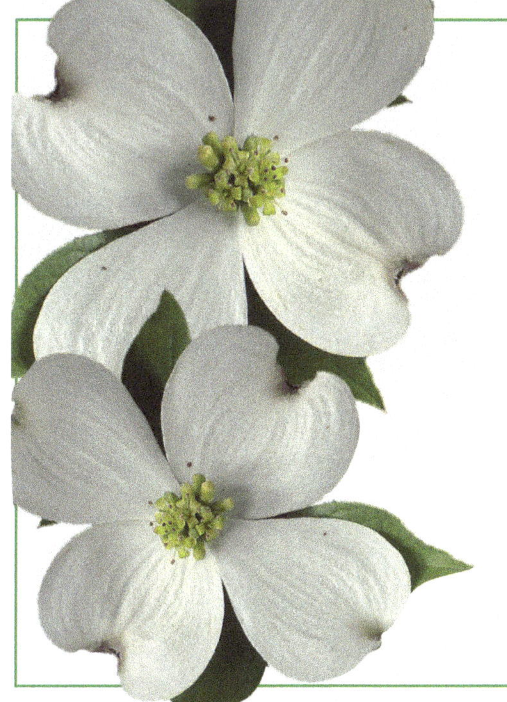

Saturday • April 2

8:00am......................Dogwood Dash
Begins at Woodville Intermediate School on Charlton Street
Sponsored by Citizens State Bank
$20 per person

9:00am-5:00pm.......Arts & Crafts Show
Downtown Woodville
Sponsored by Tyler County Texas Business Women

9:00am-1:00pm.......Variety Touring Society Motorcycle Exhibition & Antique Car Exhibition
South Side of Courthouse

2:00pm......................Dogwood Queen's Parade
Downtown Woodville

7:30pm......................Queen's Coronation and Historical Play
Fireworks to follow!
Dogwood Ampitheater • Reserved Seating
Tickets Available at the Gate: $8.00 & $10.00 each
Advance Tickets: $7.00 & $9.00 each. Available at Sullivan's Hardware,
Heritage Village, and Dogwood Pharmacy.

FOR GENERAL INFORMATION CALL THE TYLER COUNTY CHAMBER OF COMMERCE AT 409-283-2632

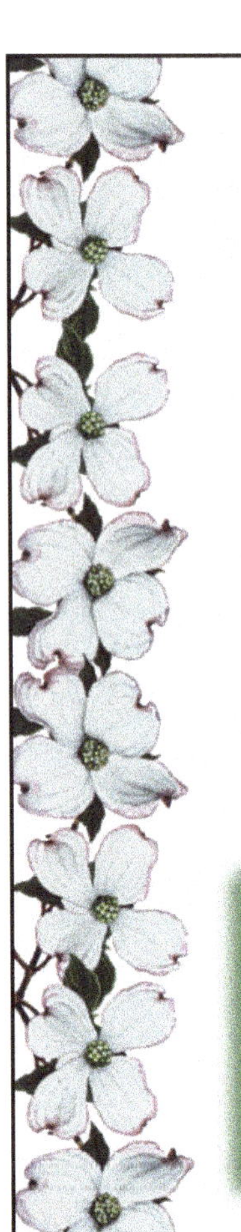

Compliments of

Buck & Donna
HUDSON
Brianna & Savannah

Welcome to Dogwood Time in Tyler County!

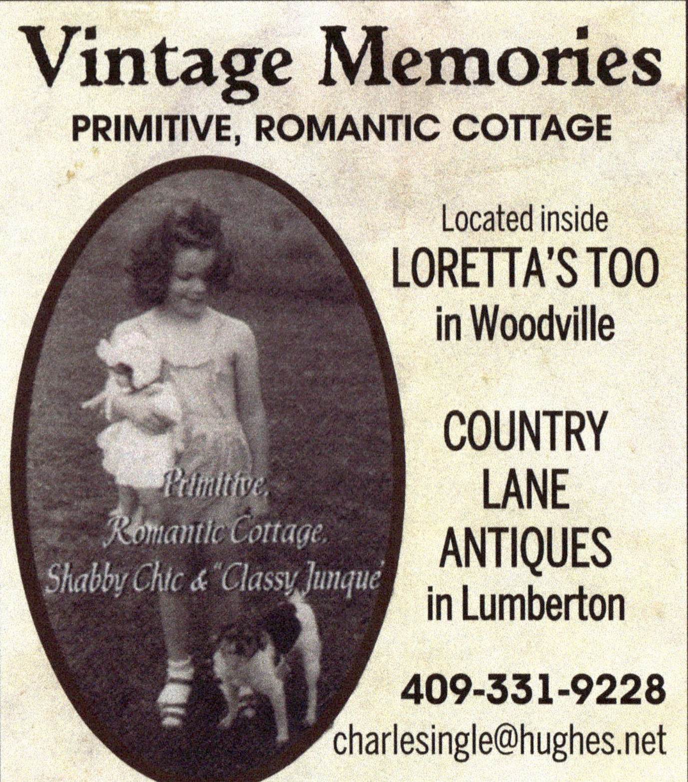

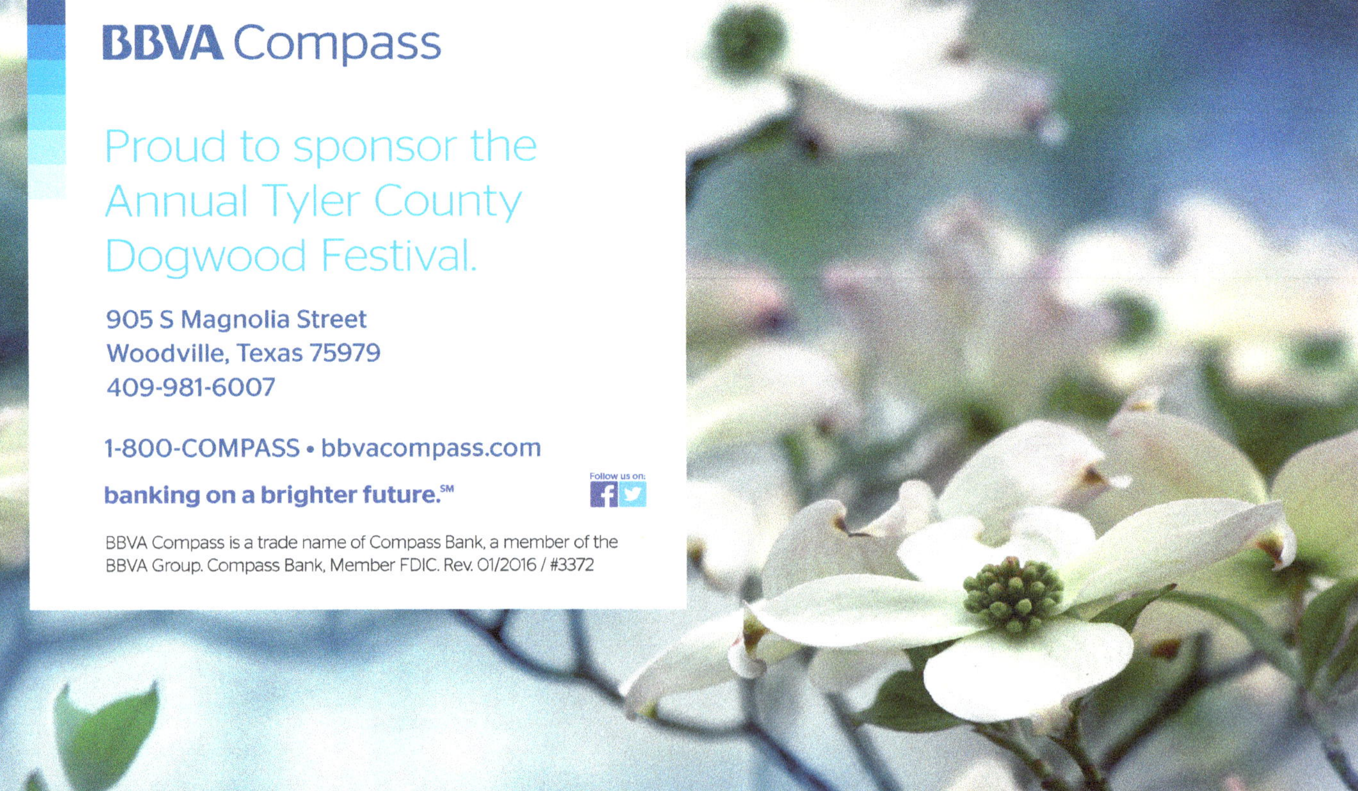

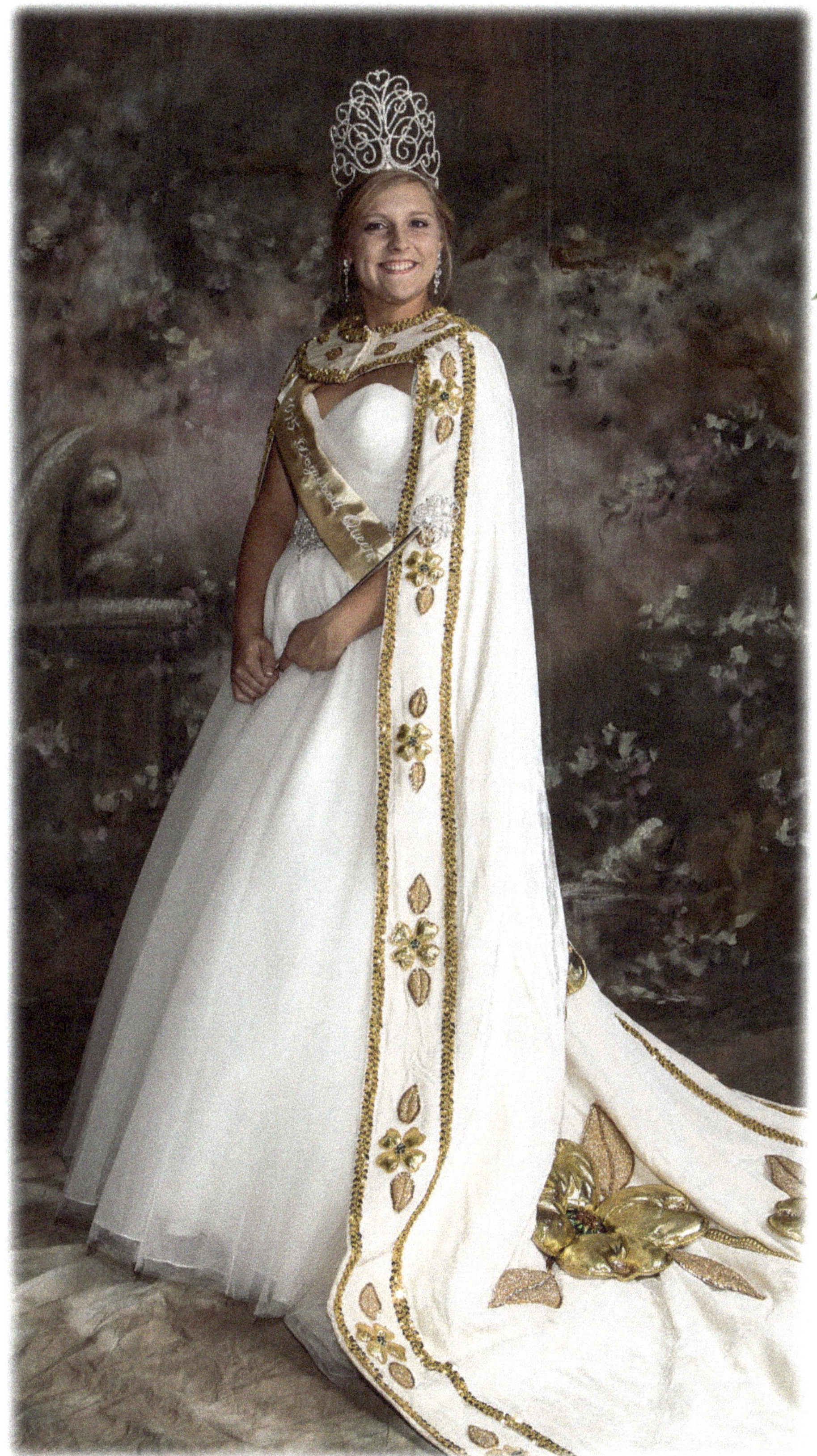

The Dogwood Festival Queen is chosen by the Kingsmen, a group of businessmen from Tyler County. Each contestant is judged on poise, beauty and personality through a series of personal interviews with the Kingsmen. The first runner-up to the queen is named the Princess of Tyler County.

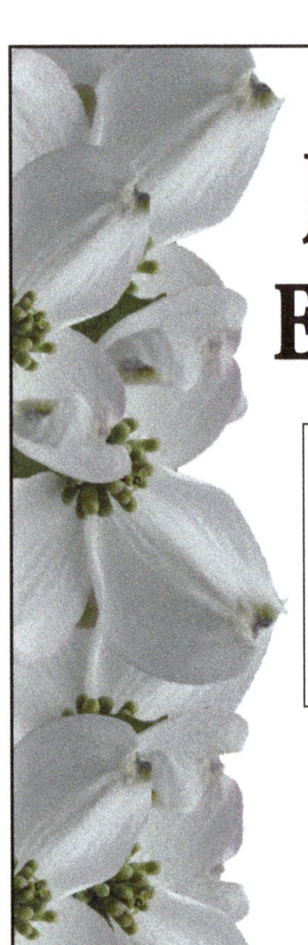

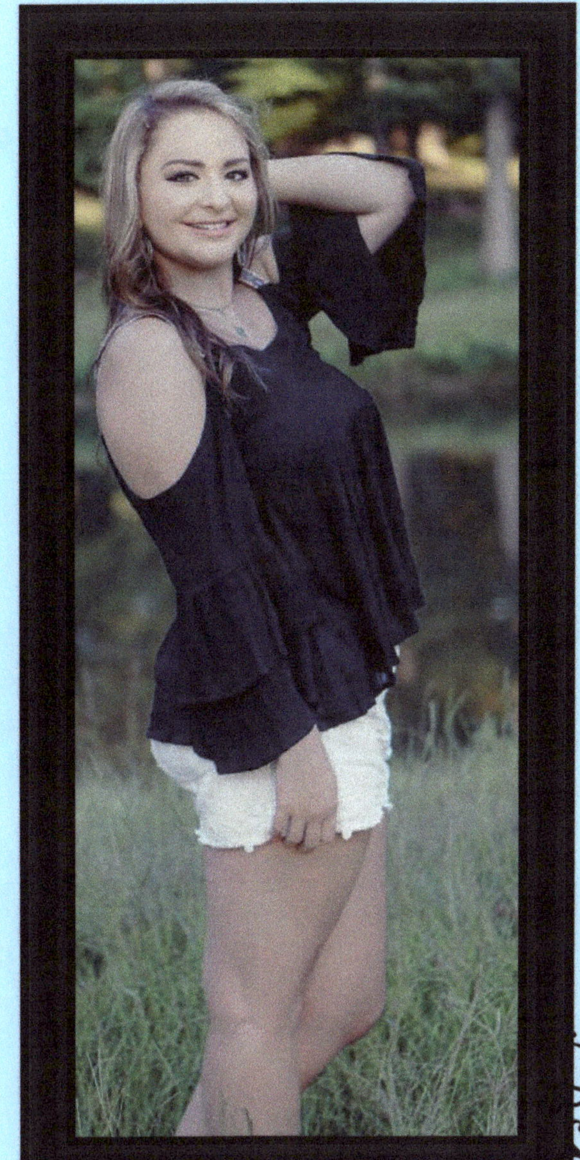

Royal Duchess of Carrollton
MADDIE HONEYCUTT

Daughter of 1989 Dogwood Queen
TANIA CELESTINE HONEYCUTT

We are so very proud
for Maddie to continue
the Dogwood tradition
and have the opportunity
to make her own

Dogwood Memories
ENJOY EVERY MINUTE!

We Love You More
Miss Maddie!

Dad, Mom,
Jordan, and
Winn Dixie

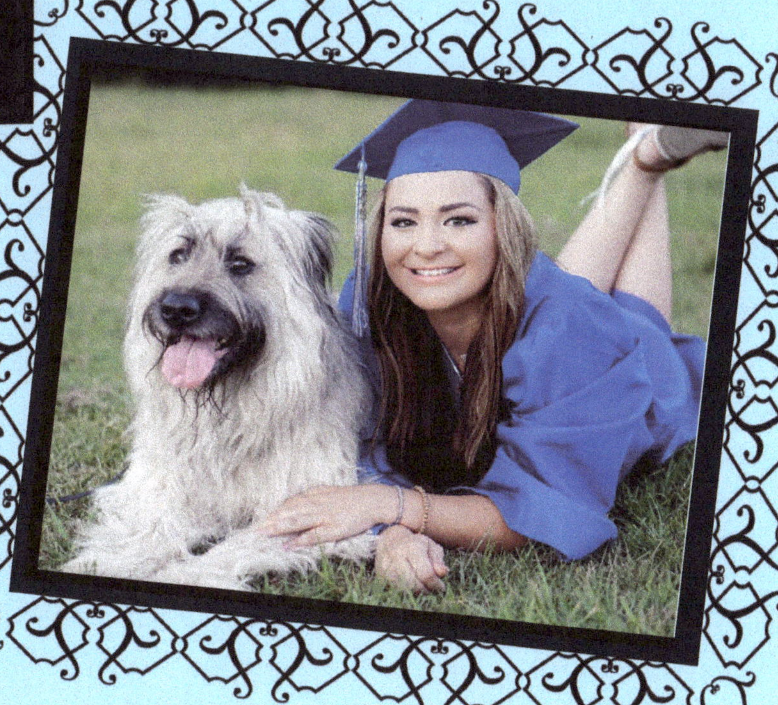

2016 PRINCESSES

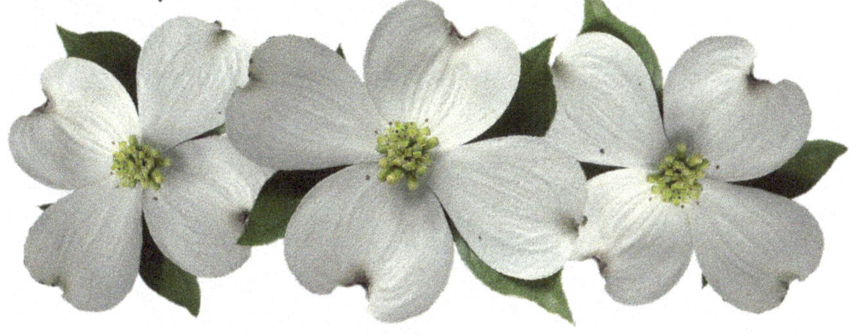

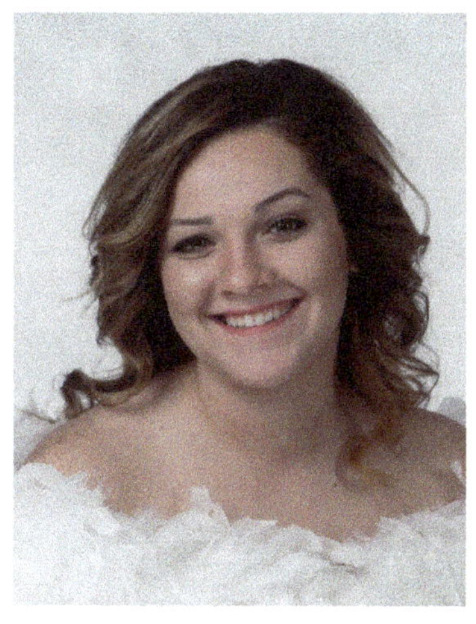

Jaci Davis
Daughter of Wayne & Lori Davis

Chester

ESCORT: BILLY THOMPSON
Son of Marty & Suzanne Thompson

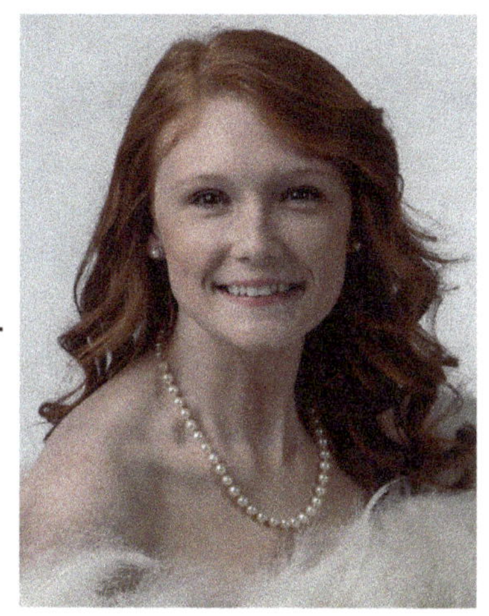

Shelby Tally
Daughter of JB & Rhonda Tally

Colmesneil

ESCORT: JEFFERSON GLENN BROWN
Son of Shawna Brown

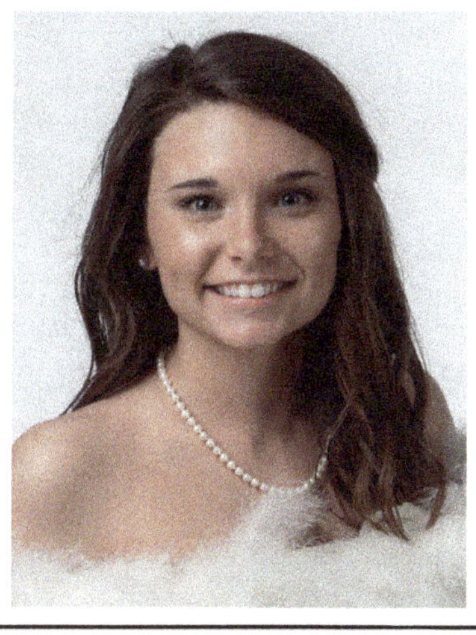

Courtney Crain
Daughter of Chuck & Alicia Crain

Spurger

ESCORT: ROBBY CLARK
Son of Betty Keller

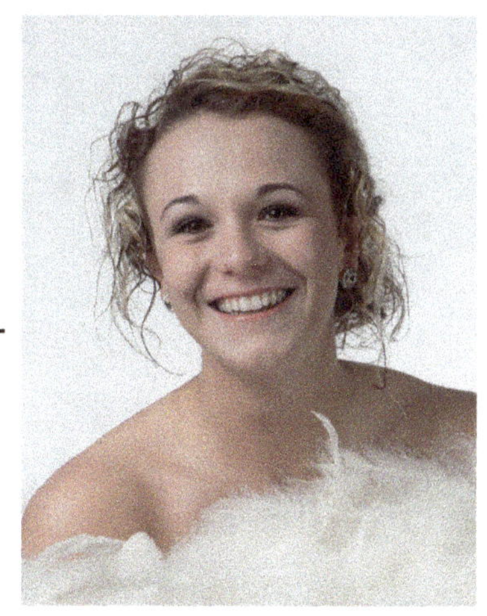

Ashlynn Whisneant
Daughter of Steven & Cathy Whisneant

Warren

ESCORT: CONNER READ
Son of Billie & Stephanie Read

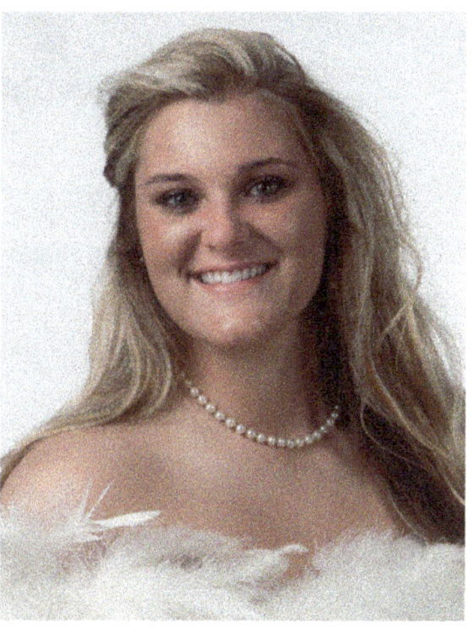

Mika Maxwell
Daughter of John & Tori Harris and Zack Maxwell

Woodville

ESCORT: DRAKE ALLISON
Son of Trey & Sharice Allison

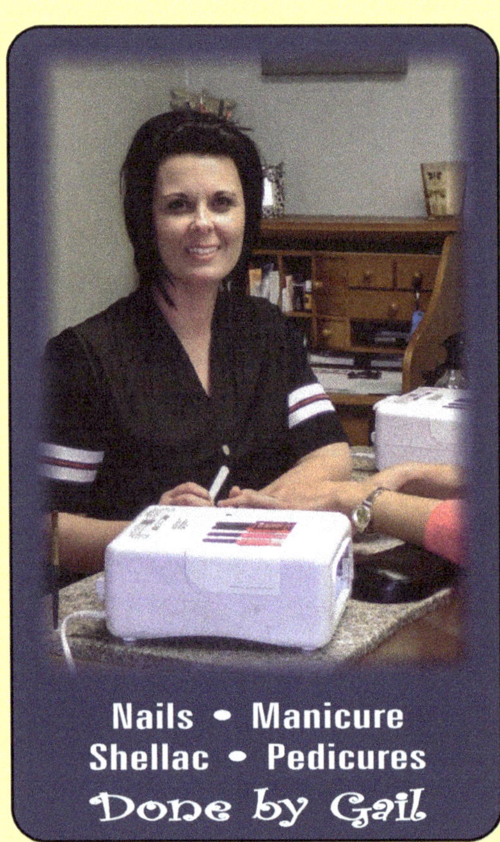

Gail's Nail Salon

BOUTIQUE

1008 West Bluff • Woodville

409-200-0042

*Enter as a Guest
and Leave as Our Friend*

YELLOWBOX SHOES, PURSES, WALLETS, WALL CROSSES, BAGS, CUSTOM T-SHIRTS, CLOTHING & MORE

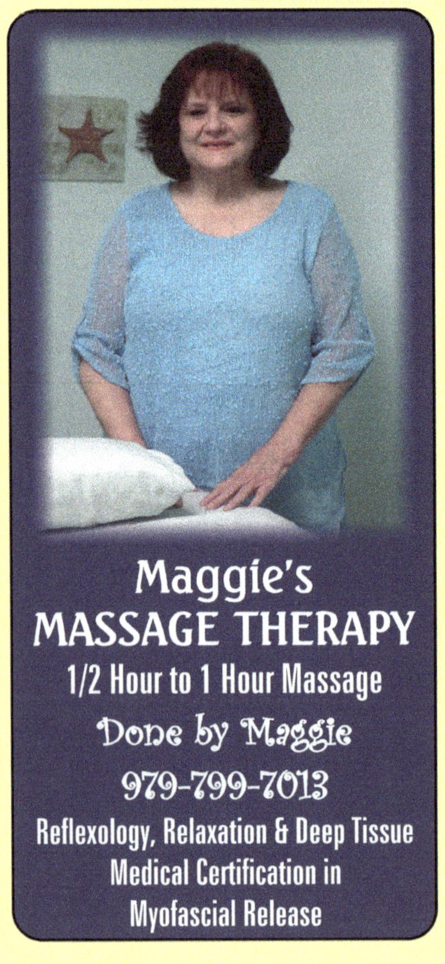

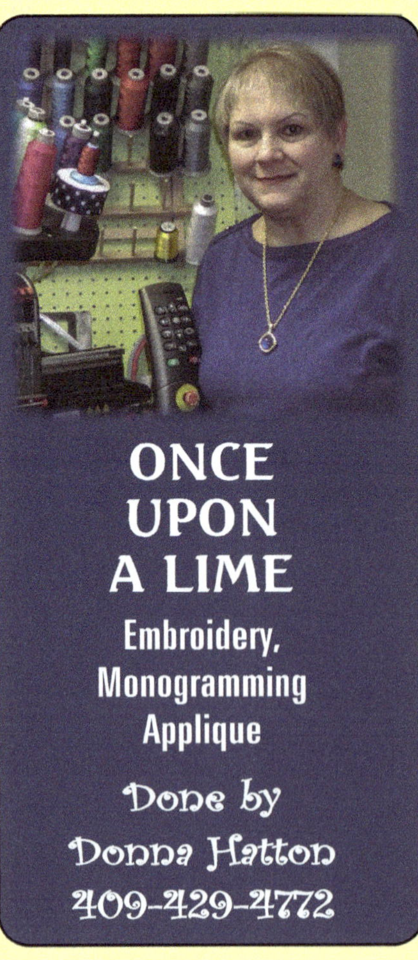

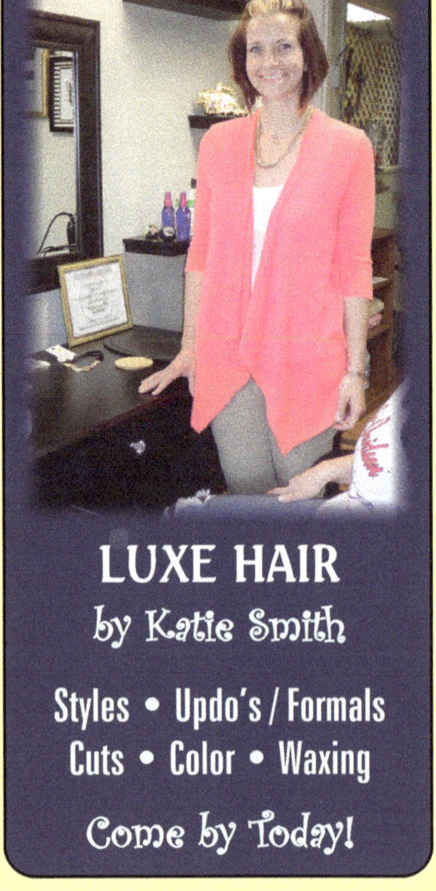

2016 LADIES IN WAITING

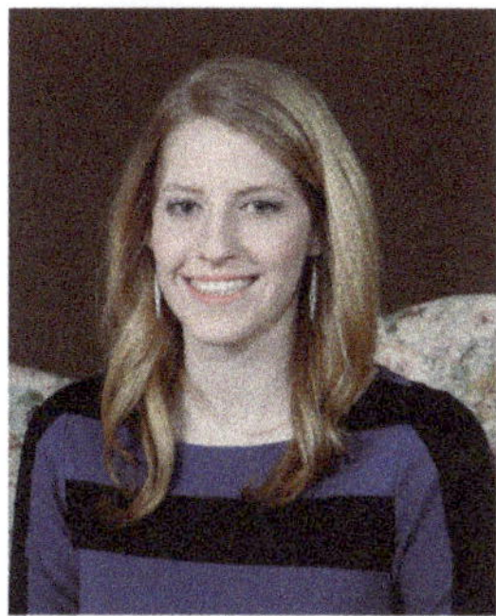

Madelaine Read
Daughter of
Wade & Stacy Read

Chester

Escort: Jared Greer
Son of James & Christy Greer

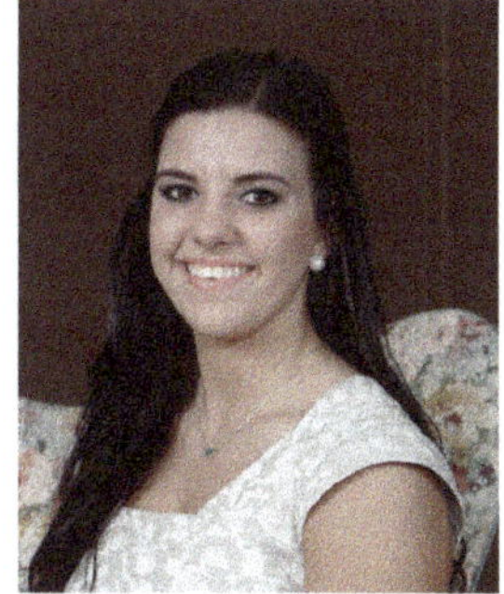

Merrick Graham
Daughter of
Lester & Penny Graham

Colmesneil

Escort: Cole Johnson
Son of Kate & Buck Johnson

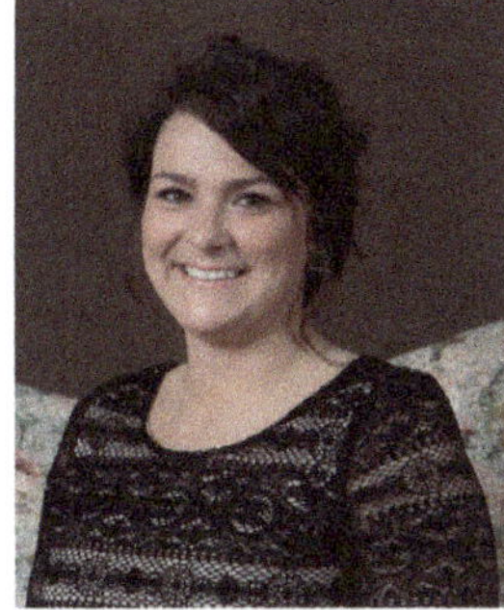

Destiney Kirkindoll
Daughter of
Billy & Gail Snider

Colmesneil

Escort: Kaleb Lindsey
Son of Chris & Shonda Lindsey

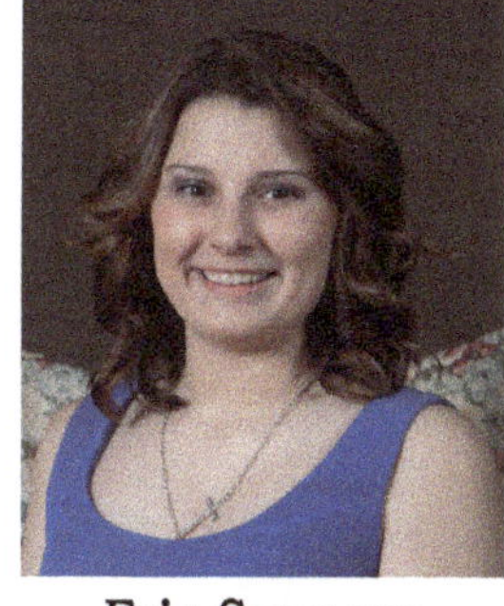

Erin Seamans
Daughter of
Lance and Heather Seamans

Colmesneil

Escort: Hunter Morgan
Son of Martie Morgan & Jerry Nugent

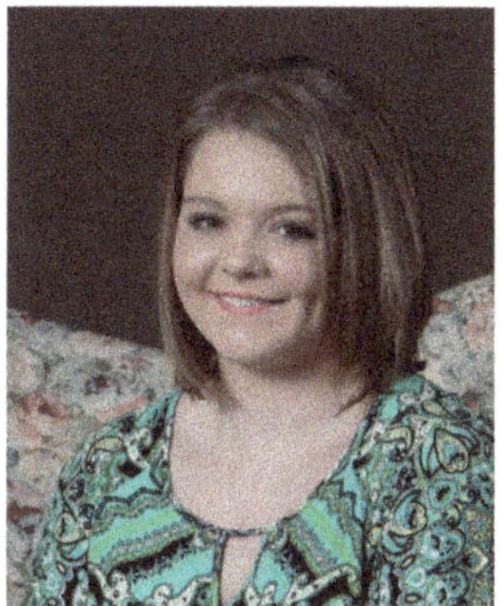

Kassondra Korthals
Daughter of
Rebekah Starr

Spurger

Escort: Randall Marcum
Son of Brent & Kristie Marcum

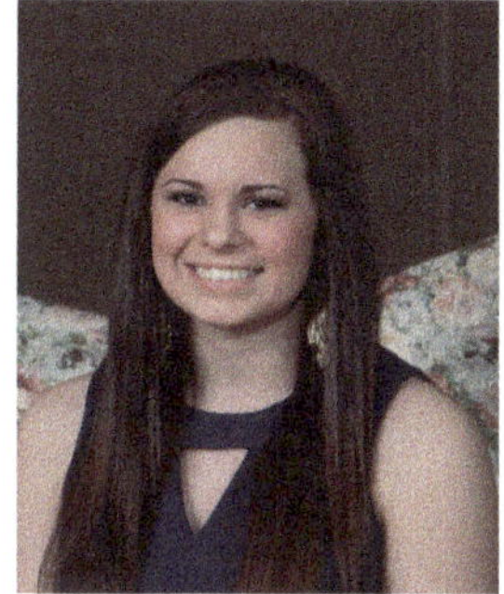

Karley Wood
Daughter of
Rickey & Mandy Wood

Spurger

Escort: Jordan Buckner
Son of Ulysses & Janeann Buckner

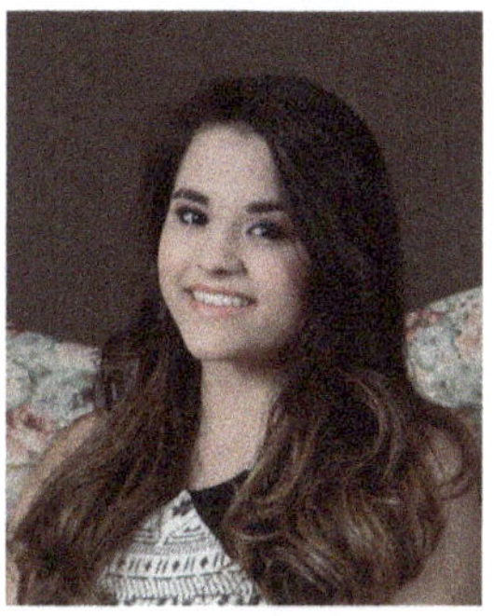

Ka'lynn Fregia
Daughter of
James & Tracy Browning

Warren

Escort: Beaux Hebert
Son of Mark & Nicole McCollister

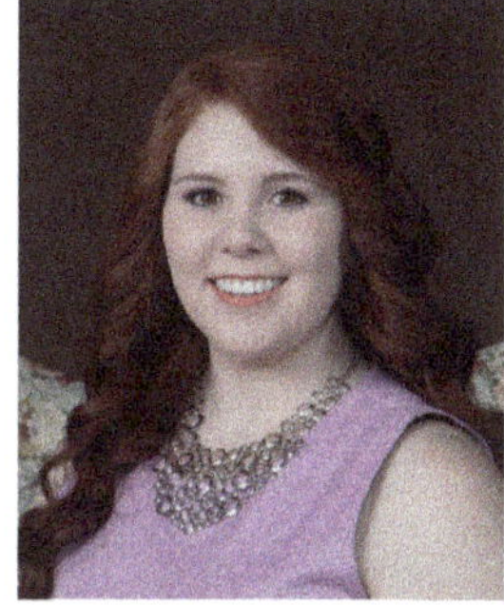

Bethany Terrell
Daughter of
James & Relinda Terrell

Warren

Escort: Tony Lyday
Son of Danyal & Dianna Lyday

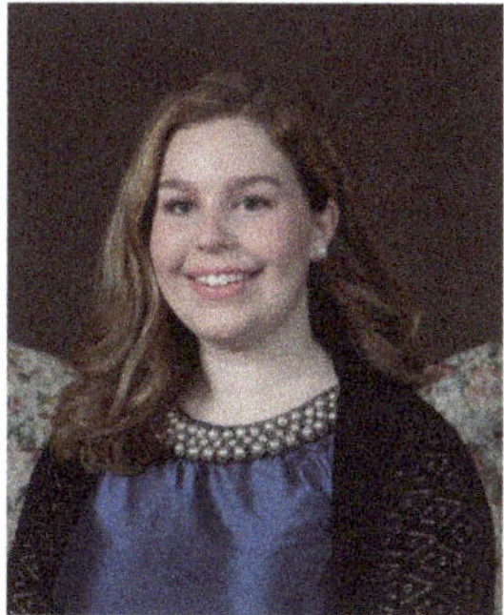

Claire Kenesson
Daughter of
Jay & Elizabeth Kenesson

Woodville

Escort: Justin Livingston
Son of John & Mandy Livingston

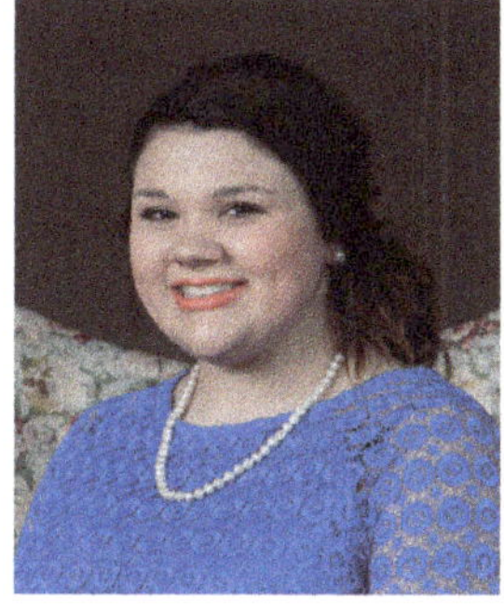

Peyton Scroggins
Daughter of
Tim & Kelli Scroggins

Woodville

Escort: Jacob Williams
Son of Lynn & Kim Williams

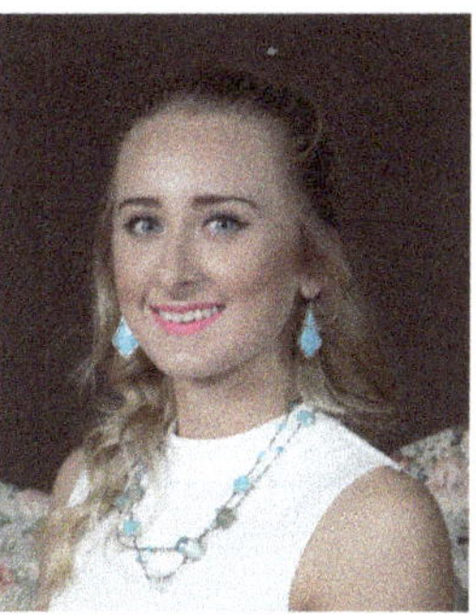

Avery Tolar
Daughter of
Casey & Neslie Tolar

Woodville

Escort: Micah Wagner
Son of Max & Shelly Clamon

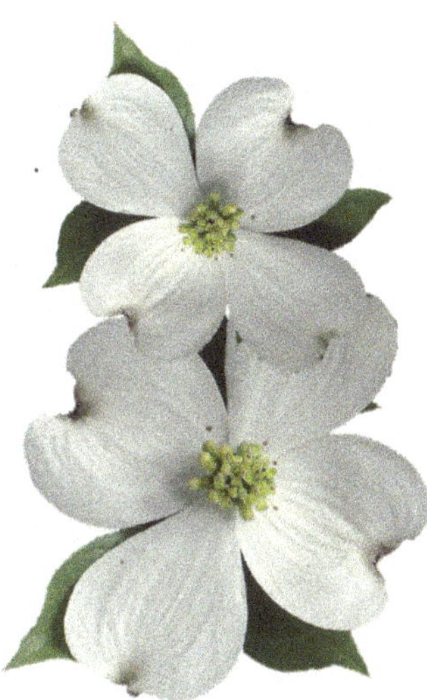

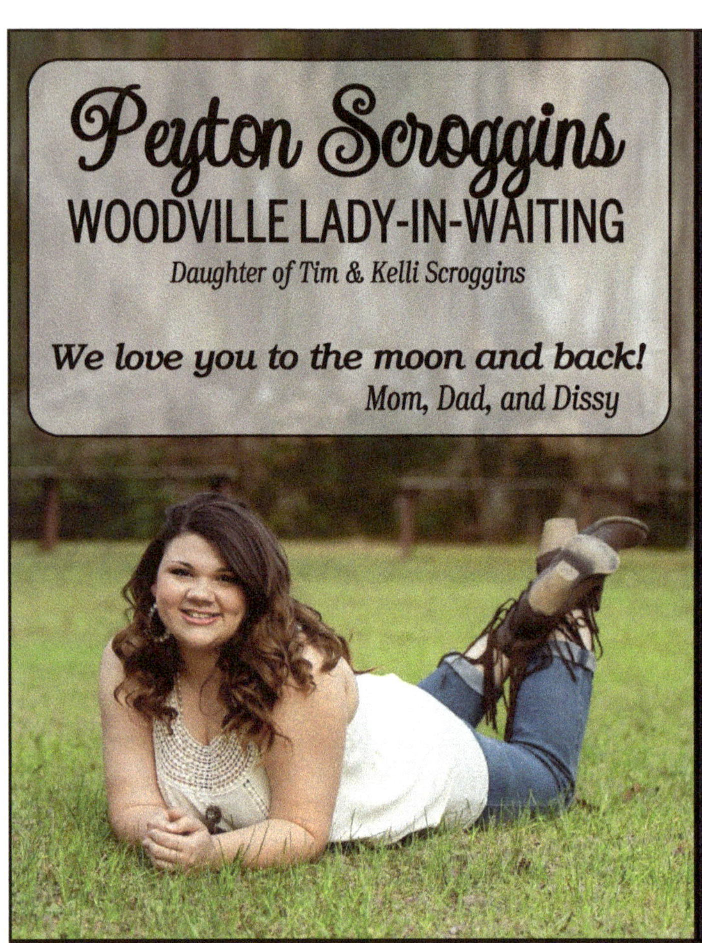

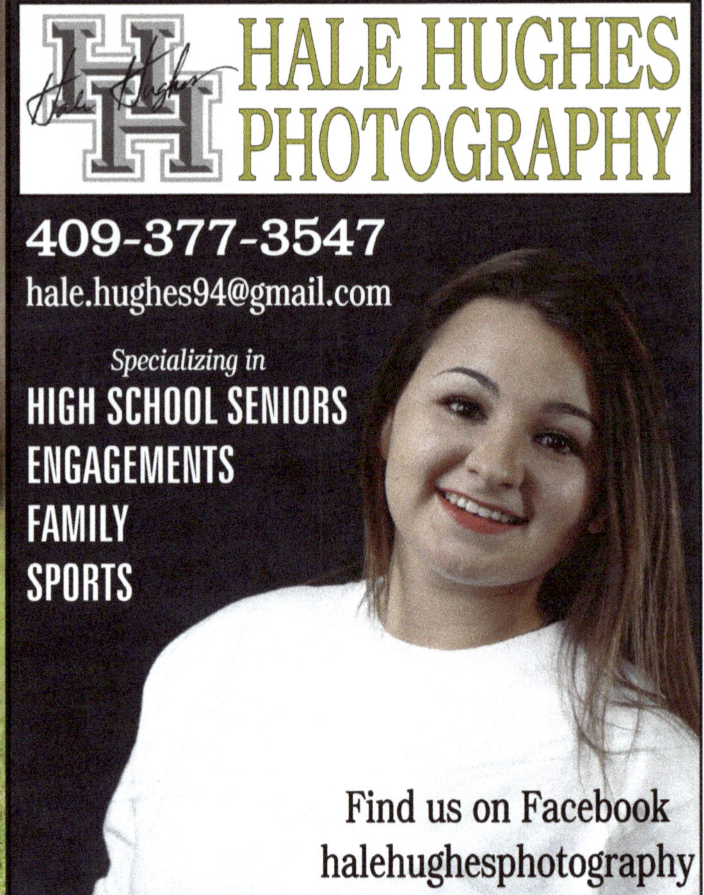

First National Bank and Dogwoods,
an East Texas Tradition.

The tradition continues as we

celebrate the 73rd annual

Tyler County Dogwood Festival.

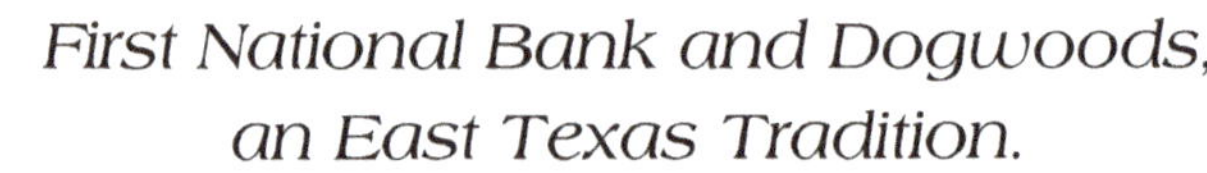

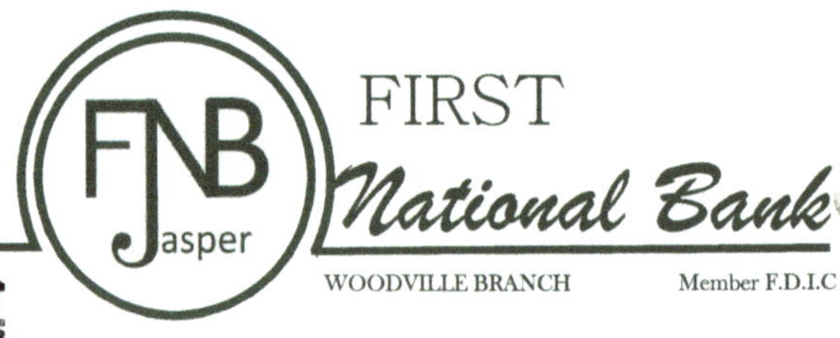

EQUAL HOUSING LENDER

270 US HWY 190 E., Woodville, Texas 75979
409-283-2831

www.fnbjasper.com

2016 TRAINBEARERS

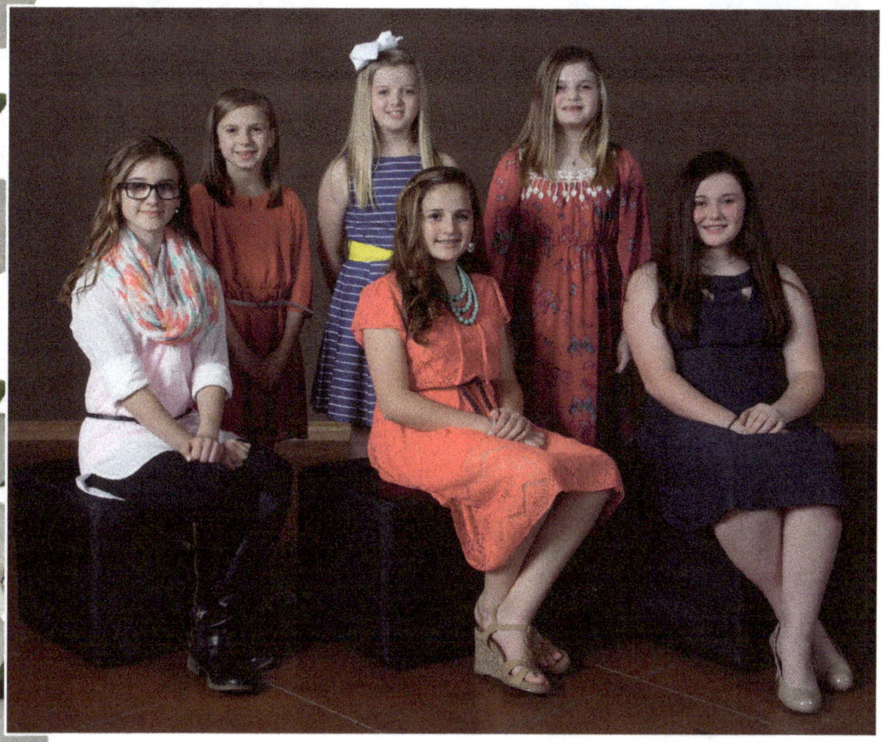

Front Row Left to Right: Sadie Calhoon, Kaela Hassig, Emma Grimes
Back Row Left to Right: Kassidy Haralson, Mollie Kate Jarrott, Virginia Boe

Emma Lee Grimes
Daughter of Bronson & Susan Grimes
Chester

Sadie Calhoon
Daughter of Donald & Melanie Calhoon
Colmesneil

Kassidy Haralson
Daughter of Bryan & Tiffany Collier
And Joshua & Jenny Haralson
Spurger

Virginia Kaitlyn Boe
Daughter of Kristen Elliott and Trey Boe
Warren

Kaela Lea Hassig
Daughter of Drew & Michelle Tucker
And Brad & Rosy Hassig
Woodville

Mollie Kate Jarrott
Daughter of Cody & Joanna Jarrott
Woodville

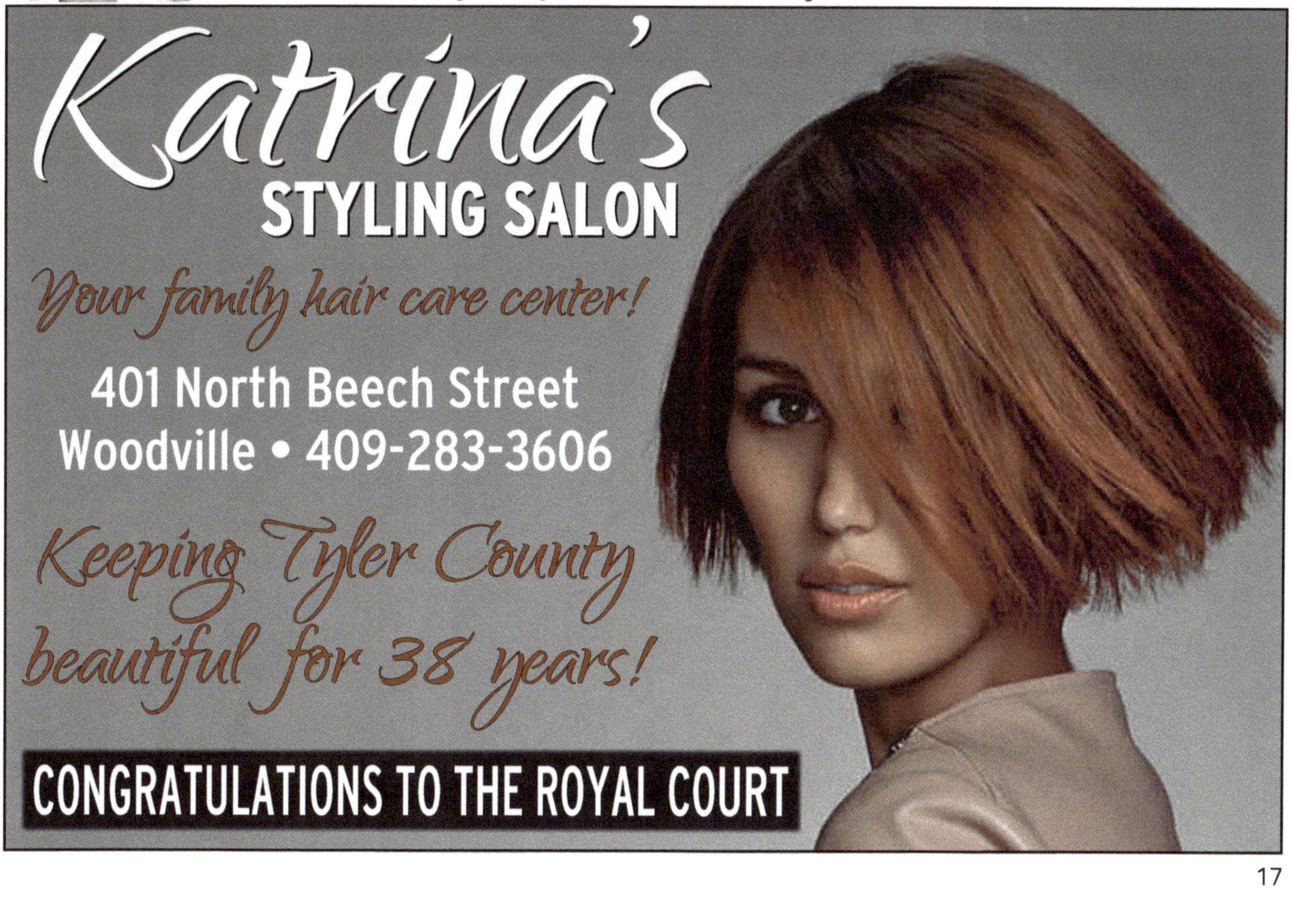

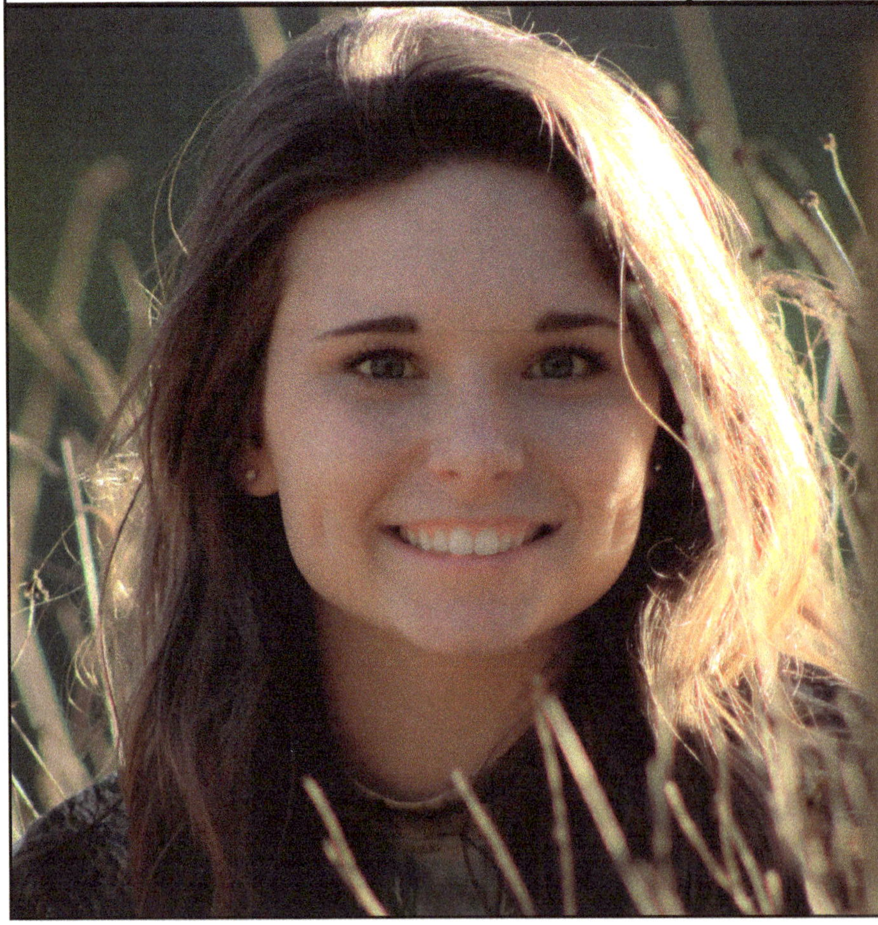

2016 CHILDREN OF THE COURT

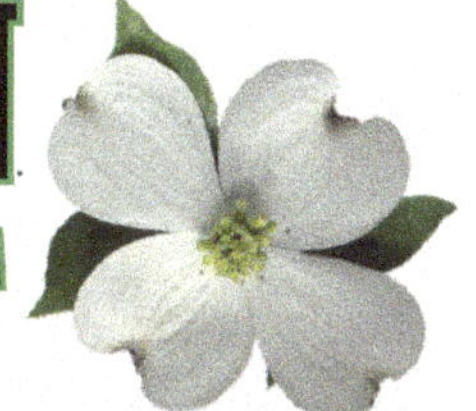

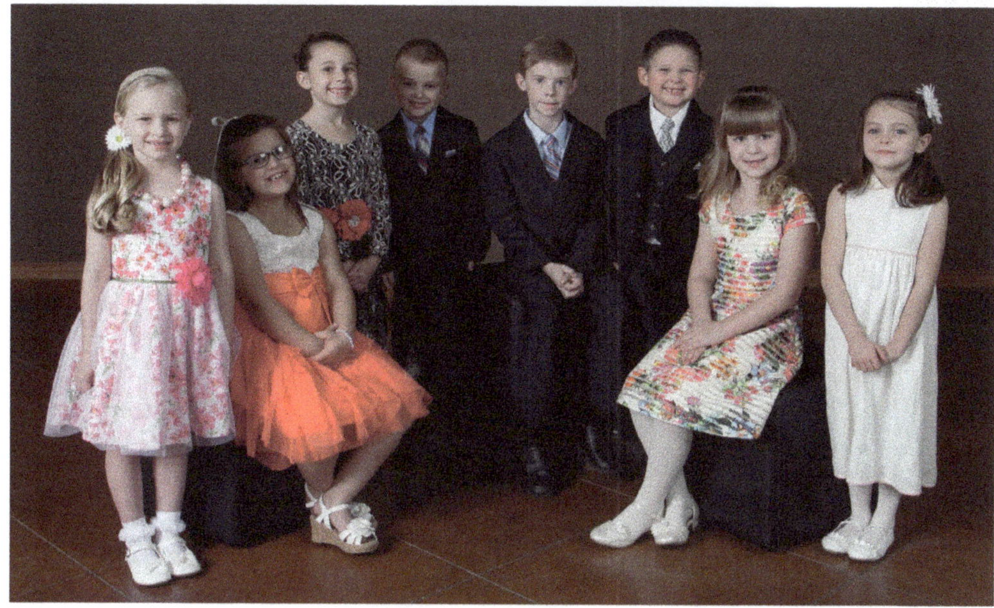

Left to Right: Kylie Winkle, Gracie Bruton, Kelsi Risinger, Carson Conner, Harrison Dillard, Jake Risinger, Allie Ogden, Julia Jeansonne.

Allie Ogden ~ Flower Girl
Daughter of Scott & Lacy Ogden
Chester

Gracie Bruton ~ Flower Girl
Daughter of Courtney McHenry
Colmesneil

Kylie Winkle ~ Flower Girl
Daughter of Nicole Blankenship
Spurger

Harrison Dillard ~ Crown Bearer
Daughter of Robby & Kelly Dillard
Warren

**Kelsi Risinger ~ Flower Girl
and Jake Risinger ~ Ring Bearer**
Children of John David & Mandy Risinger
Woodville

Julia Jeansonne ~ Flower Girl
Daughter of Fulton & Sharon Jeansonne
Woodville

Carson Conner ~ Scepter Bearer
Son of Joseph & Lauren Connor
Woodville

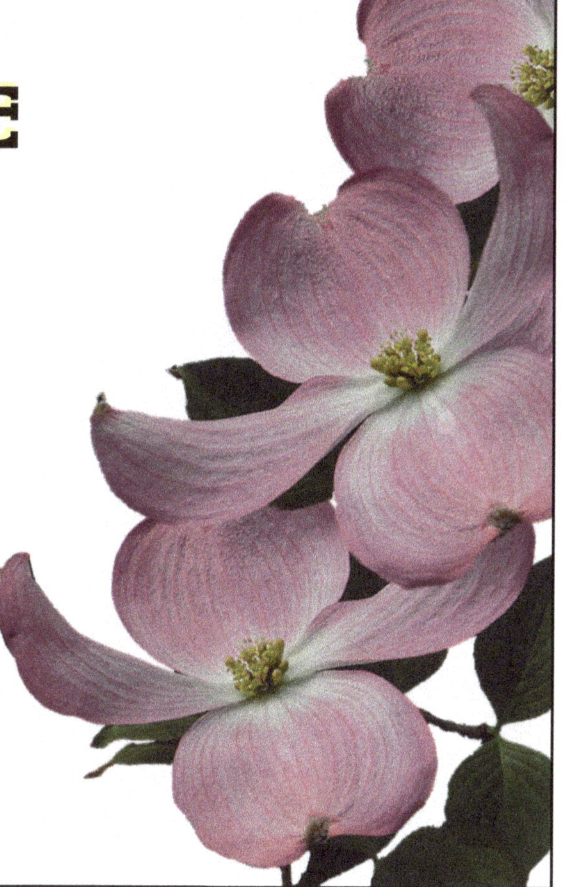

Compliments of

DIXIE & GEORGE JARROTT

Celebrating the Beauty of Springtime

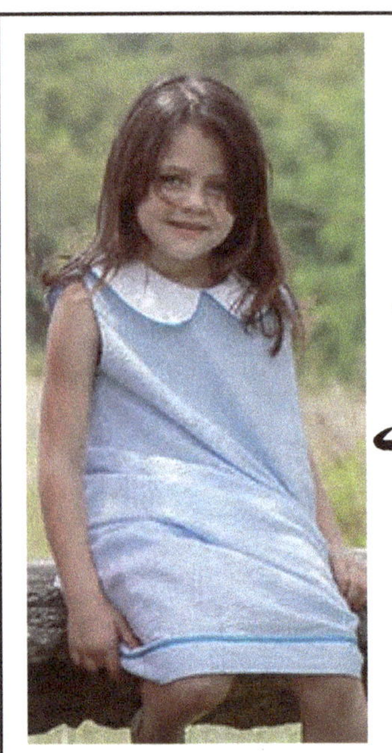

Congratulations

Julia May Jeansonne

2016 FLOWER GIRL

Love, GG and Pierre

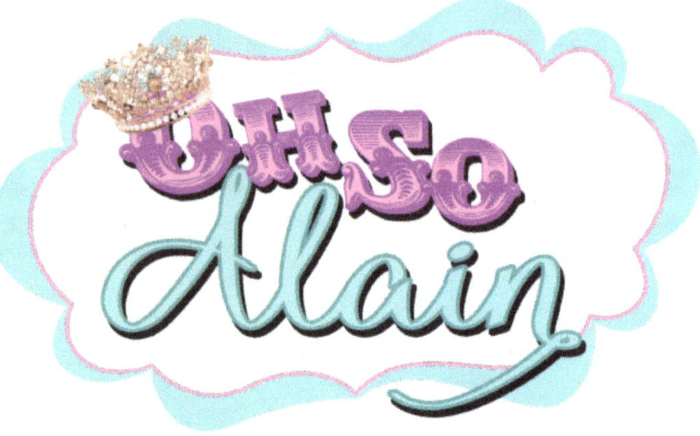

BEAUTY SALON

218 East Bluff
Woodville, TX 75979

409-377-5941

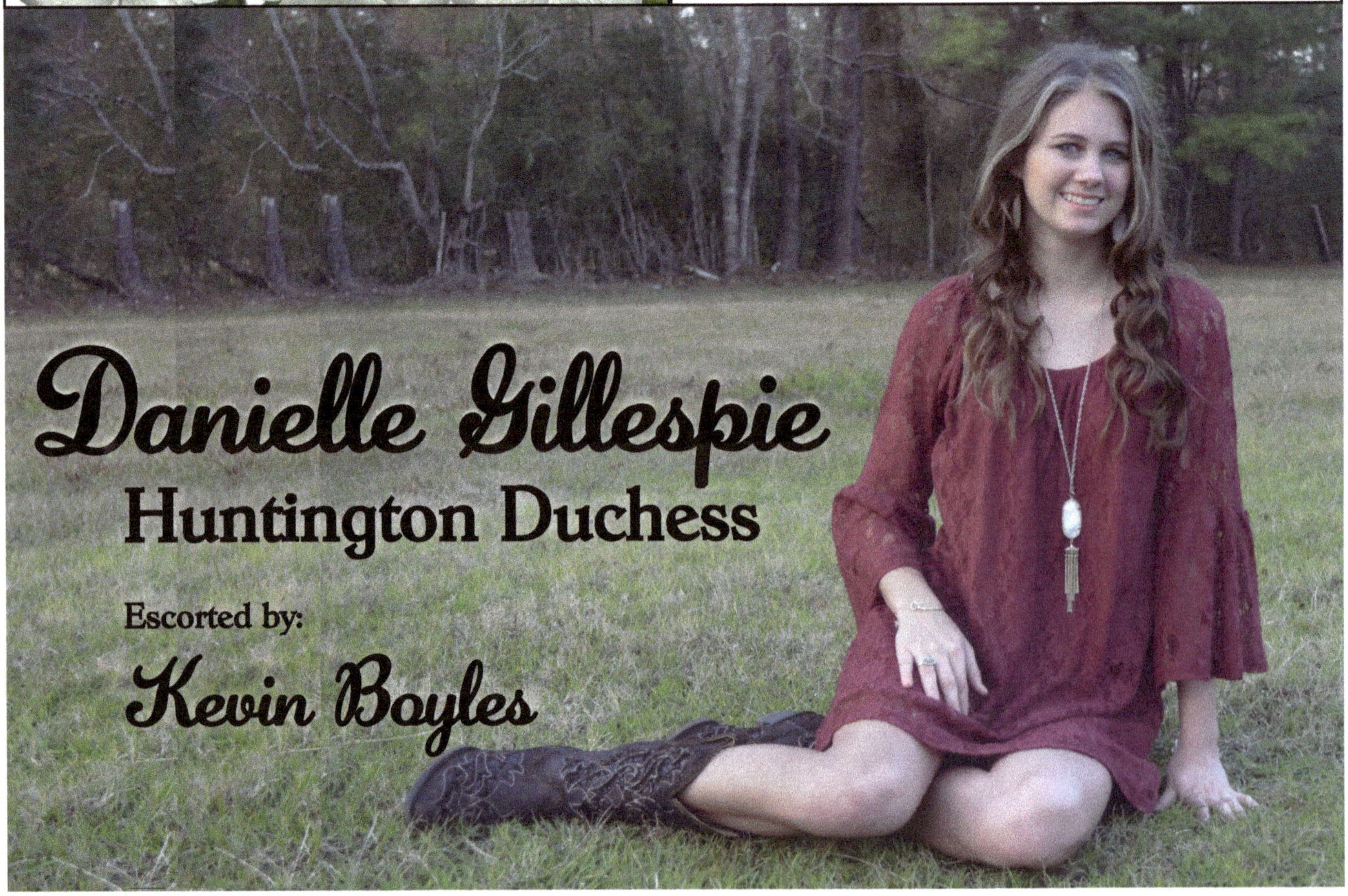

Danielle Gillespie
Huntington Duchess

Escorted by:

Kevin Boyles

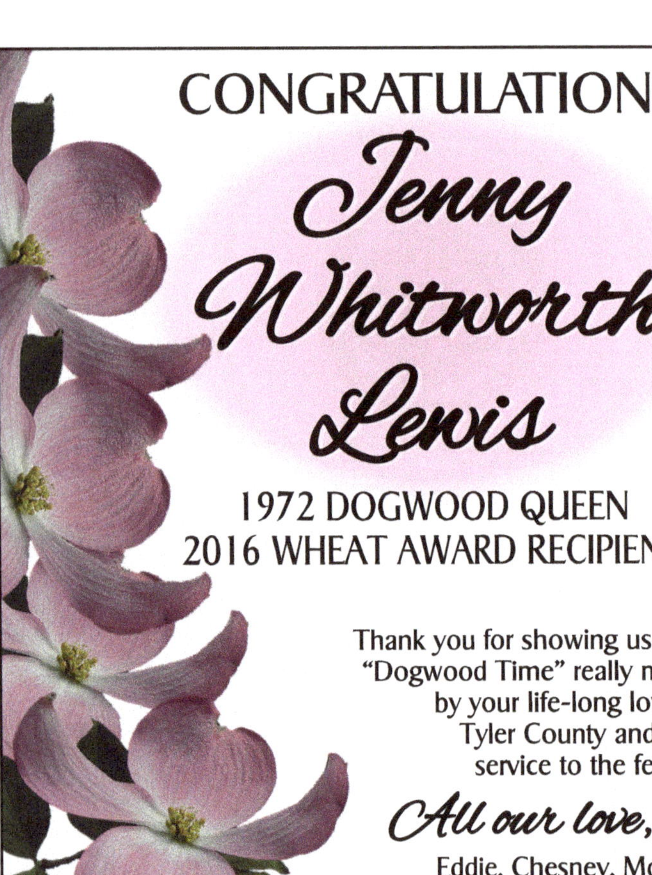

CONGRATULATIONS
Jenny
Whitworth
Lewis
1972 DOGWOOD QUEEN
2016 WHEAT AWARD RECIPIENT

Thank you for showing us what
"Dogwood Time" really means
by your life-long love for
Tyler County and your
service to the festival.

All our love,
Eddie, Chesney, Morgan,
Rick, Michelle,
Kirby, and Carter

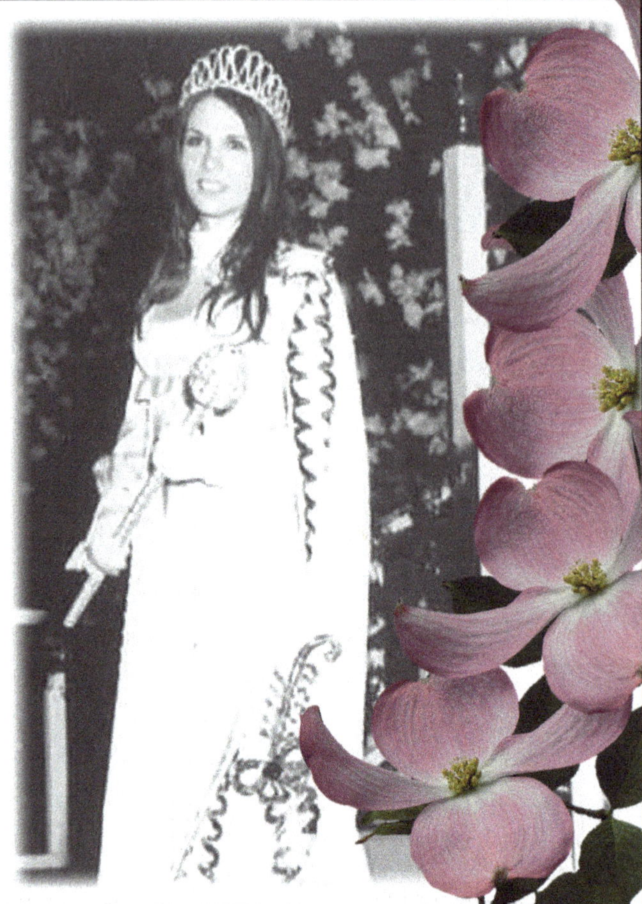

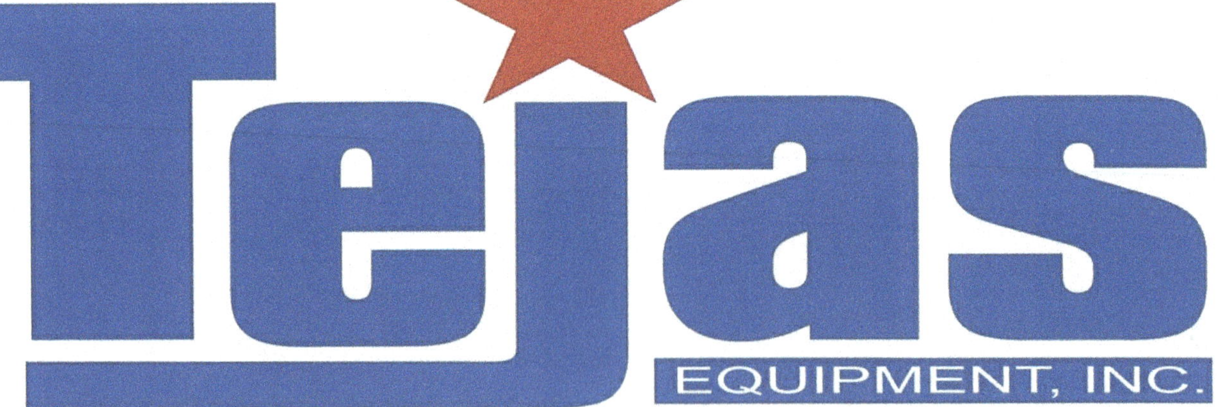

SALES, PARTS & SERVICE
TeJas
EQUIPMENT, INC.
HILLISTER
409-331-0111
LUFKIN
936-639-1009
www.tejasequipmentinc.com

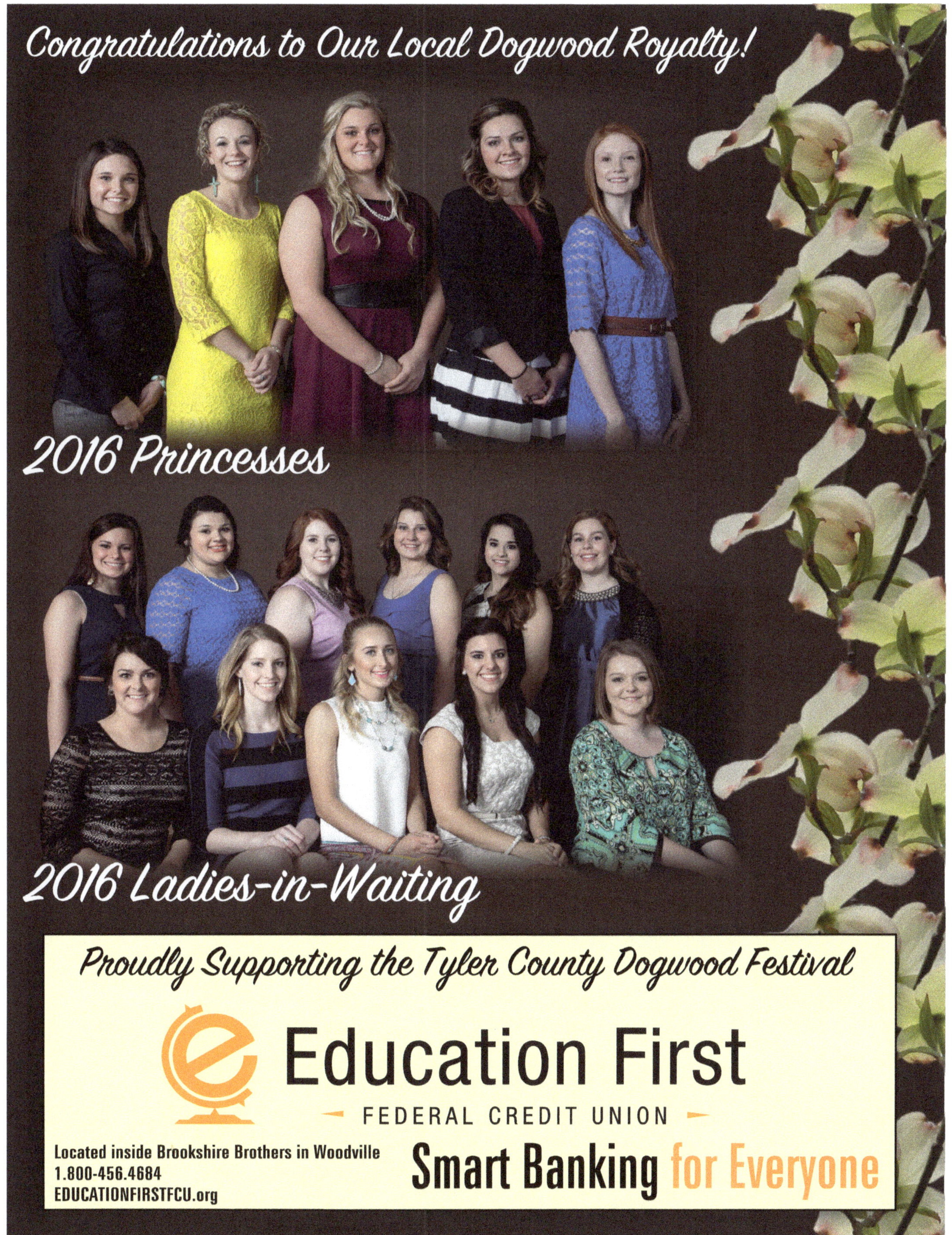
Congratulations to Our Local Dogwood Royalty!

2016 Princesses

2016 Ladies-in-Waiting

Proudly Supporting the Tyler County Dogwood Festival

Education First
FEDERAL CREDIT UNION
Smart Banking for Everyone

Located inside Brookshire Brothers in Woodville
1.800.456.4684
EDUCATIONFIRSTFCU.org

1967
O. R. Crawford
General Manager of
Wothwestern Timber
Company

1968
Glenn E. Richard
Chairman of the Board,
Gulf States Utilities Co.

1969
**Arleigh Brantley
Templeton**
President of
Sam Houston
State University

1970
Arthur Temple
President of
Temple Eastex Corporation

1971
**Allan
& Marialice
Shivers**
Former Govenor
& First Lady of Texas

1972
John Gray
President of Lamar University

1973
Edward Clark
Former Ambassador
to Australia & Native East Texan

1974
Charles Wilson
U.S. Congressman
Disrict 2

1975
Ralph Ramos
Journalist & Author

1976
Ralph Steen
President of
Stephen F. Austin
State University

1977
Price Daniel
Former Texas
Supreme Court Justice
& Govenor of Texas

1978
Ben Rogers
Philanthropist & Businessman

1979
Don Adams
Former State Legislator

1980
**Herbert E.
Dishman**
Business Leader
& Philanthropist

1981
Joe J. Fisher
U.S. District Judge
Eastern District Judge

1982
**Francis E.
Abernathy**
Author, Folklorist
& Professor of English

1983
**C. L. "Charlie"
Schmucker**
Teacher, Banker
& College Administrator

1984
Roy D. Shotts
Radio & Television
Pioneer in Port Arthur

1985
Louie Welch
Businessman &
Former Mayor of Houston

1986
Bill Hobby
Lieutenant Govenor of Texas

1987
Maury Meyers
Mayor of Beaumont

1988
**James Henry
"Red" Duke, M.D.**
TV Personality,
Surgeon & Professor

1989
**Charles T.
Terrell**
Chairman of the
Texas Criminal Justice
Task Force

1992
Allen Hightower, Jr.
Texas House of Representatives,
District 18

1993
**Dr. James
Anderson Allums**
Thoracic and
Cardiovascular Surgeon

1994
Walter Diggles
Executive Director of
Deep East Texas
Council of Government

1995
Larry Beaulieu
Vice President & General Manager
of KFDM-TV in Beaumont

1996
Horace McQueen
Broadcaster of Farm & Ranch
News on KLTV-TV in Tyler and
KTRE-TV in Lufkin

1997
Mike Moses
Texas Commissioner
of Education

1998
Walter Umphrey
Texas Attorney

1999
James B. Hull
Director, State Forestor,
Assistant Vice Chancellor for
Agriculture Programs

2000
Tom Harken
Restaurant Owner,
Business Leader, Speaker & Writer

2001
**Joe Wesley
Dickerson, M.D.**
Medical Doctor & Founder of
Dickerson Memorial Hospital
in Jasper

2002
**Dr. James M.
Simmons**
President of
Lamar University

2003
Lester Lowery
Retired Sawmill and
Chipmill Owner

2004
Evelyn Lord
Mayor of Beaumont

2005
Nick Lampson
Former Congressman

2006
Howard Rosser
President of Rosser
Communication Inc
& Executive Director of
East Texas Tourism Association

2008
Joe Folk
Retired Teacher, Superintendent,
and Jasper County Judge

2009
**Chief Oscola
Clayton M.
Sylestine**
Principal Chief of the
Alabama-Coushatta Tribe of Texas

2011
Jim McReynolds
Texas House of Representatives,
District 12

2013
Mic Cowart
Entergy Customer Service Manager
for Tyler & Hardin Counties

2014
Joe Penland
Philanthropist & Businessman

2015
David J. Waxman
Planning Consultant

The Mr. East Texas Award is presented to the East Texan who best exemplifies the spirit and quality of leadership which advances, shapes and gives direction to the growth and progress of East Texas.

Mr. East Texas
Harold Estes

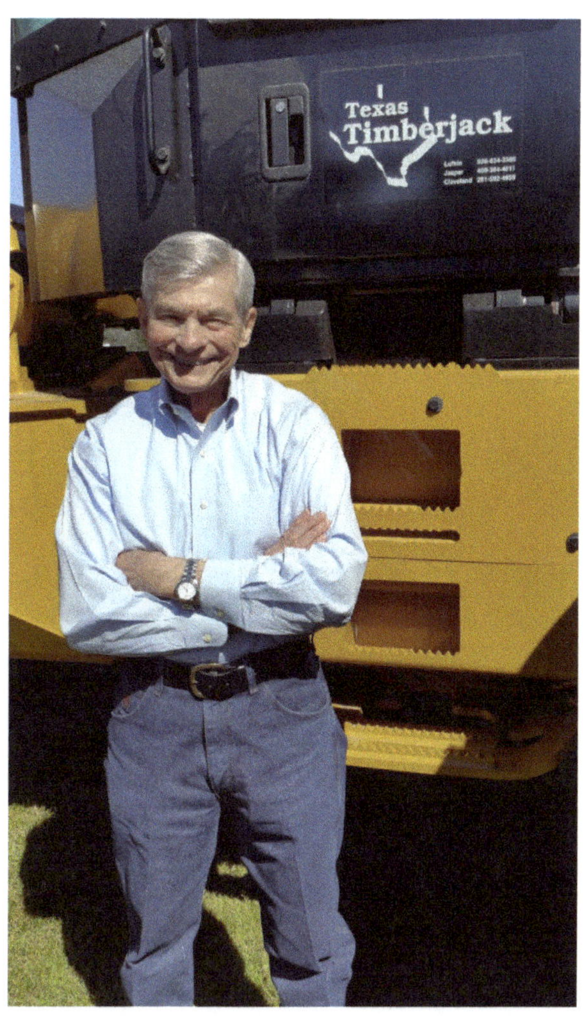

Harold Estes was born in 1940 back in the mountains of North Carolina. He grew up with 3 sisters, a stay at-home-mom and a dad who worked in construction, logging, and farming. Because of the hard winters, most of this work was seasonal. Therefore, Harold had to pitch in to help his dad. By the age of six he was working with his dad skidding logs down the mountain with a draft horse. He learned how to use a chainsaw and fall a tree right where his dad told him to. So when he thought about what he would like to do as a profession, forestry was a natural thing for him to choose. However, even with scholarships, he did not have enough money to go to school. This did not deter him from his dream. He was able to get a construction job for two summers in Greenland. With this opportunity, he was able to go to college.

He chose Mars Hill College, a small Baptist two year school, tucked away in the mountains near Asheville, NC. He was pretty excited and highly motivated because his dream of going into the logging business was actually taking legs. From Mars Hill he transferred to the University of Georgia where he majored in Forestry in 1964. He later went back to school at Georgia State and took a number of business classes.

While he was at Mars Hill, he met his future bride of 52 years. He was not planning on getting married, but he met a girl who thought otherwise. It did take her three years to finish her degree and close the marriage deal. He was ready now to get started on his logging business dream. One needs money to do that, but the "hounds of poverty" were still chasing him. The closest he could get to the logging business was to work as a forester and then a Timberjack salesman.

After a variety of jobs at Timberjack, he was transferred to Lufkin as manager of their factory store. In 1984 he bought the factory store and renamed it Texas Timberjack. Subsequently, he has been engaged in real estate development, scrap metal, various oil and gas activities, sawmill and wood treating, and other such investments. Harold has served on several board of directors...Overhill Farms, TreeCon Resources, and Newton Bancshares.

He and his wife, Conni have two sons and three grandchildren. He enjoys spending time with his family and friends. He loves to fish in the Kenai River in Alaska and travel with the oldest grandson. And as most of you who know him, you know he loves to make a deal.

The charities supported through private funds and The Estes Education and Charitable Foundation are: Educational Scholarships in area schools, Harmony Hill Baptist Church, Mosaic Center, Museum of East Texas, Junior League, Ellen Trout Zoo, The Joseph House, Harold's House, Texas Forestry Museum, Angelina Arts Alliance, Lufkin Landscape Task Force, Marine Corps Association Foundation, Buckner Family Place, and the Pregnancy Center.

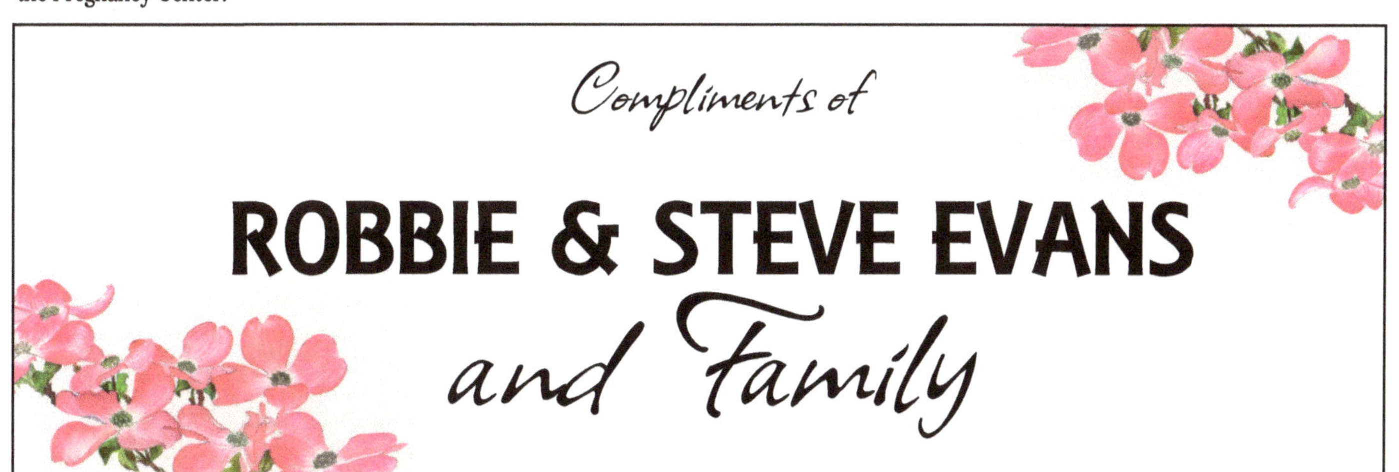

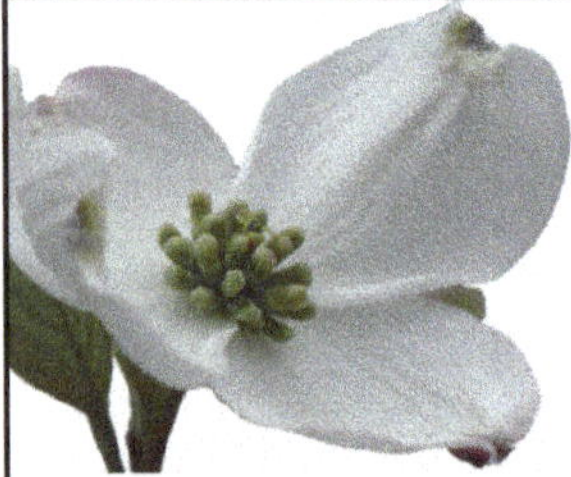

THE CITY OF COLMESNEIL

Proudly Supports Our Local Dogwood Court!

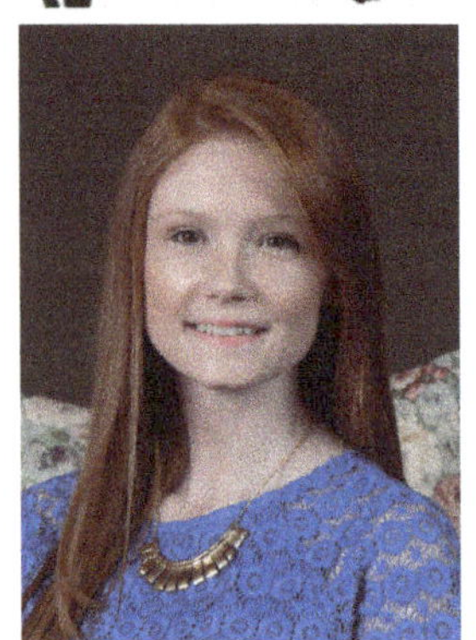

Shelby Tally
PRINCESS

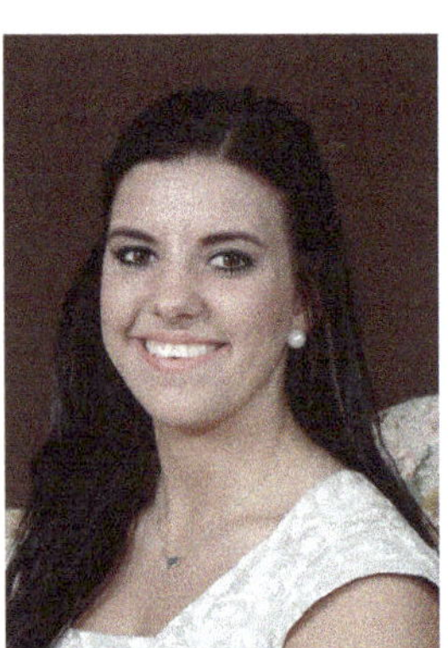

Merrick Graham
LADY-IN-WAITING

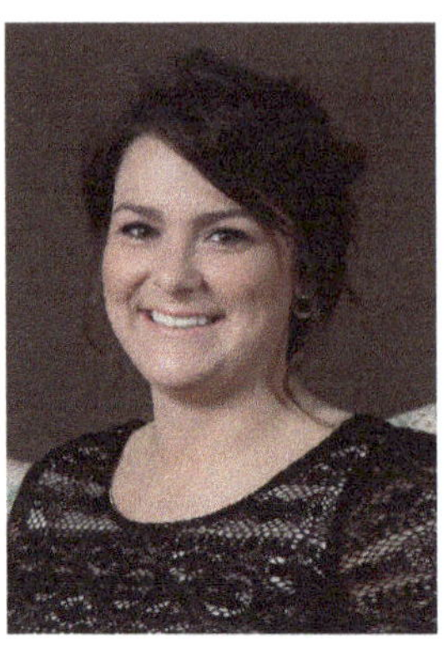

Destiney Kirkindoll
LADY-IN-WAITING

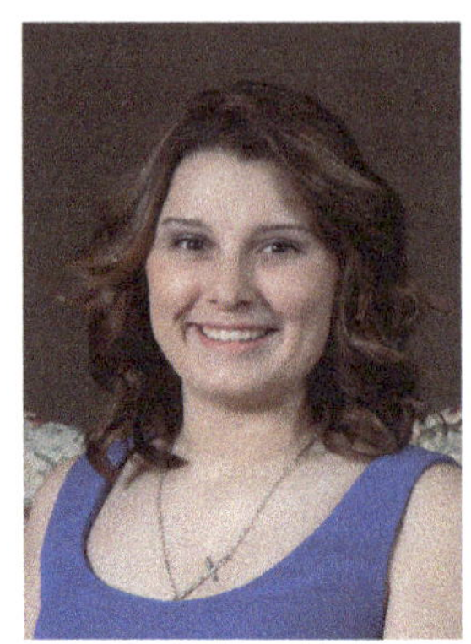

Erin Seamans
LADY-IN-WAITING

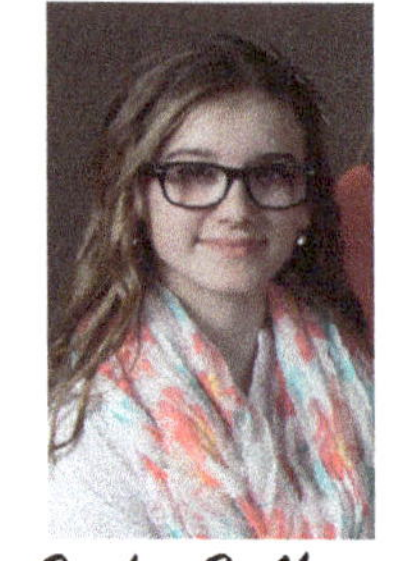

Sadie Calhoon
TRAINBEARER

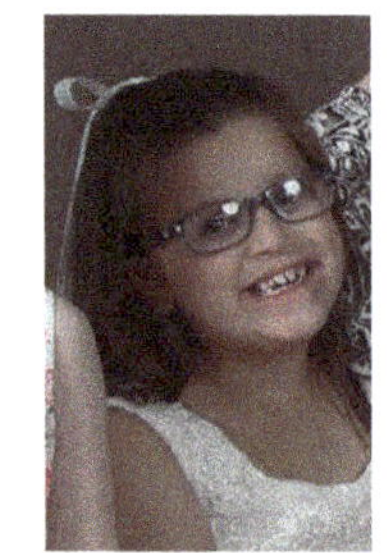

Gracie Bruton
FLOWER GIRL

CONGRATULATIONS GIRLS!

Parade Marshall
Bob Boykin

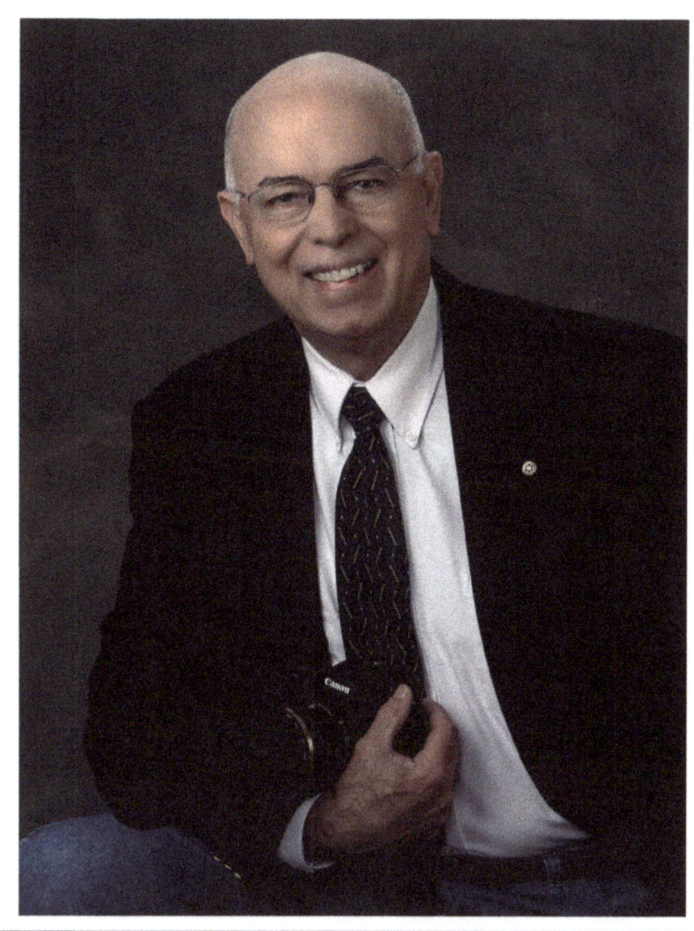

The Tyler County Dogwood Festival is pleased to announce Bob Boykin as the 2016 Parade Marshal. Bob has served as festival photographer since the 1970's. As a Tyler County native, he has a long history with the festival having participated in the play and parade activities since childhood. He began photographing festival events as a personal project while in college. In 1971, the Publications Committee began using some of his photos for the book and publicity materials.

Mr. Boykin owned and operated a professional photography studio, as well as a pharmacy in Woodville and has captured countless community events and personal memories for Tyler County residents for more than 40 years. He has served the festival as a Kingsman since the early 1970's and was awarded the James E. Wheat award in 2008.

In January 2016, Bob resigned as festival photographer and sold his studio. The festival would like to thank Bob for sharing his time and talent with us for all of these years and we wish him well as he moves on to spend more time with his wife Janie, daughters Leslie and Grace, and their families.

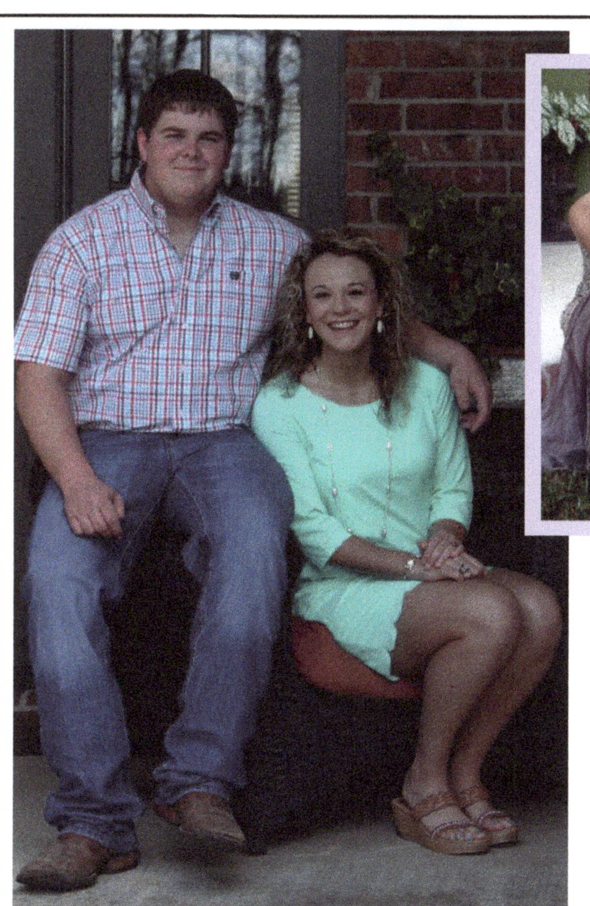

2016 VISITING DUCHESSES

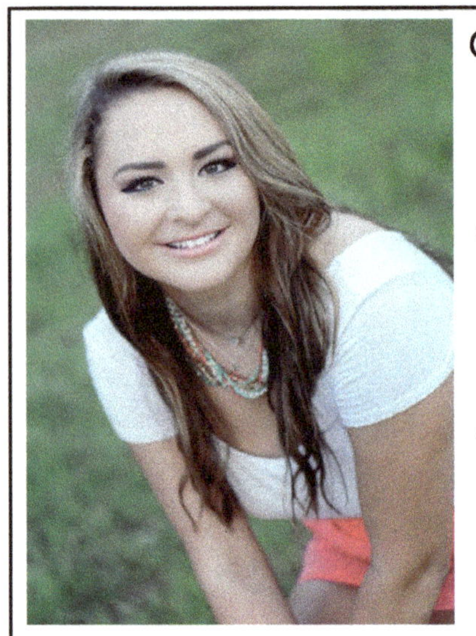

Royal Duchess
Maddie
Honeycutt
CARROLLTON
Daughter of
Craig & Tania Honeycutt

Maddie's Mother
TANIA CELESTINE
HONEYCUTT
*was the 1989
Dogwood Queen*

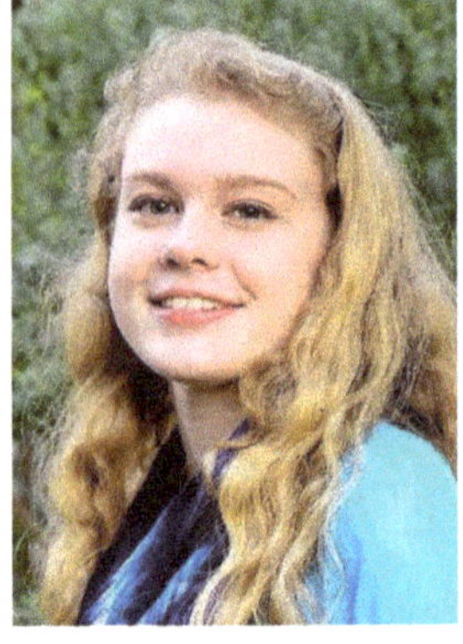

Janie Nichols
SELMA
Daughter of
Julie & Grant Nichols

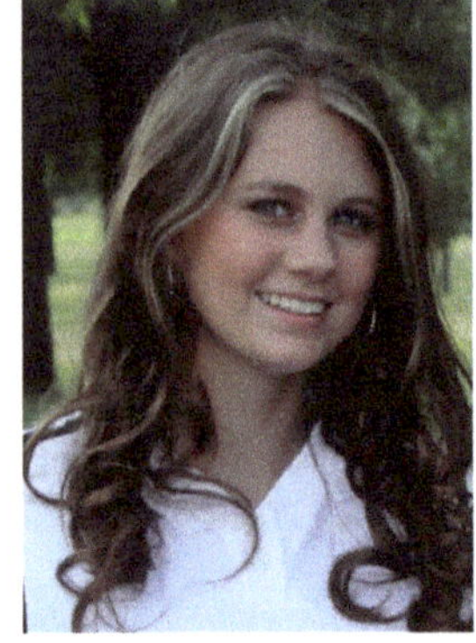

Danielle Gillespie
HUNTINGTON
Daughter of
Aaron & Sheila Gillespie

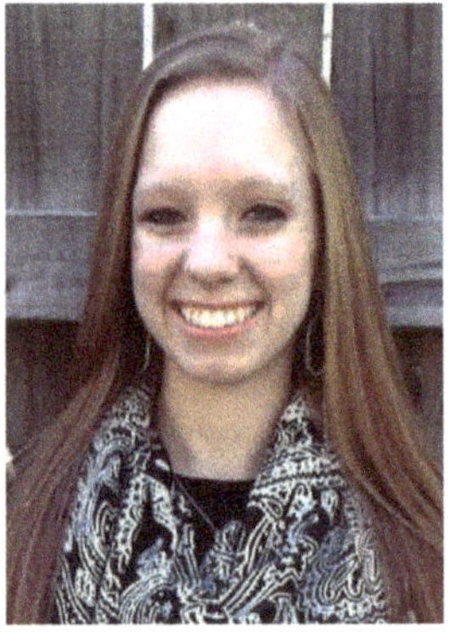

Angela Ocnaschek
BEAUMONT
Daughter of
Tim & Teri Ocnaschek

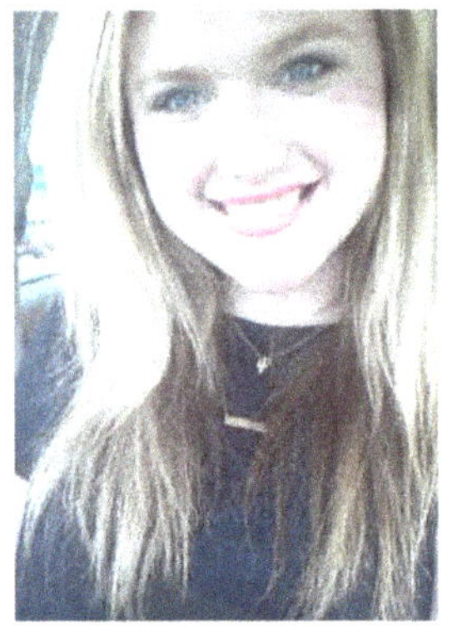

Jamie Shumake
JASPER
Daughter of
Amy & Brandon Shumake

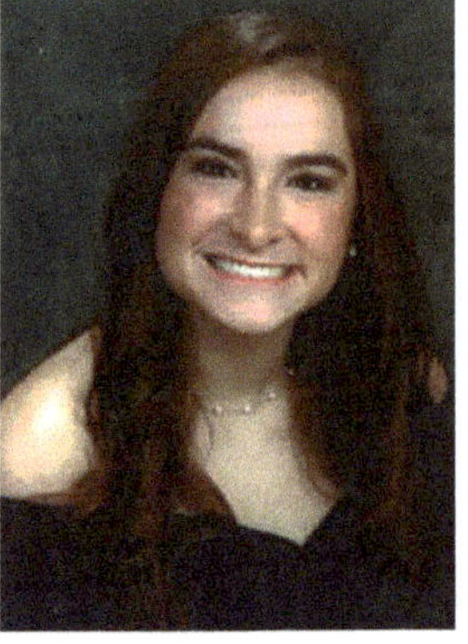

Erin Hester
BEAUMONT
Daughter of
Alan & Kimberly Hester

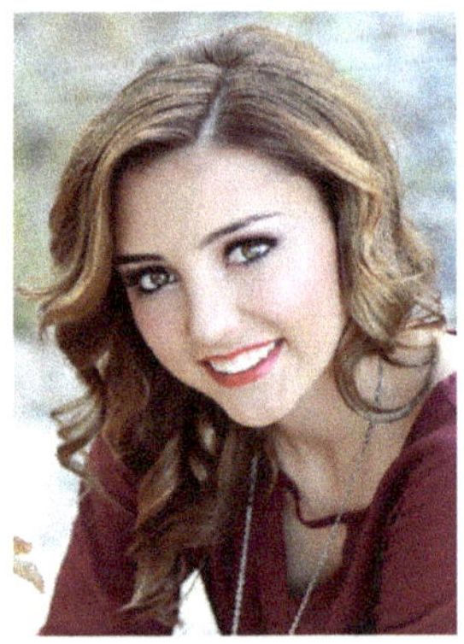

Haley Mackey
HUDSON
Daughter of
Cara Hewitt & Scott Mackey

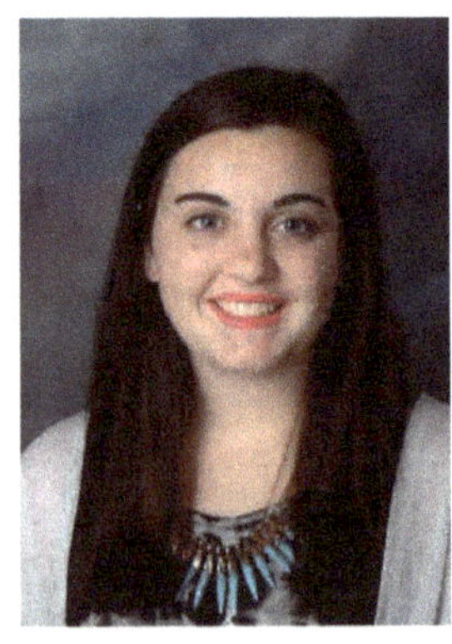

Mia Passalaqua
CONWAY, AR
Daughter of
David & Pele Passalaqua

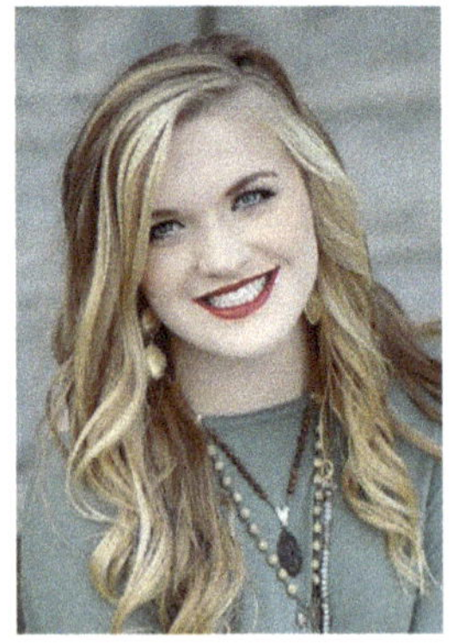

Catherine Golden
JASPER
Daughter of
Mr. & Mrs. Andrew Golden

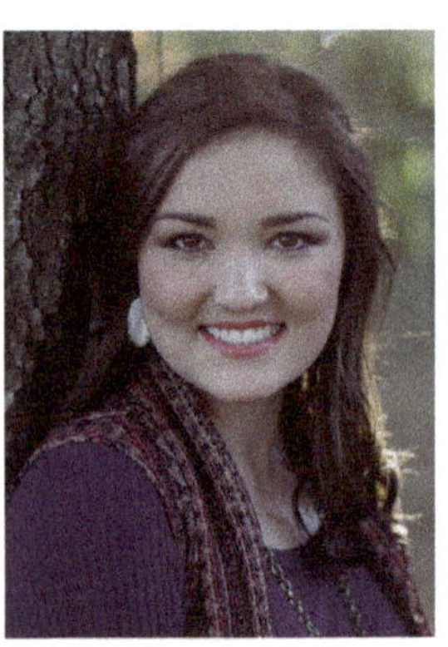

Rebecca Modisette
HUNTINGTON
Daughter of
Mike & Sonia Modisette

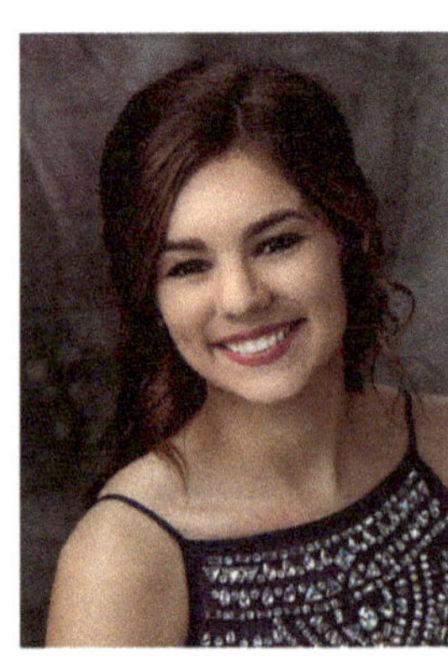

Brittni Lachausse
LONGVIEW
Daughter of
Kevin & Kelli Lachausse

2016 VISITING DUCHESSES

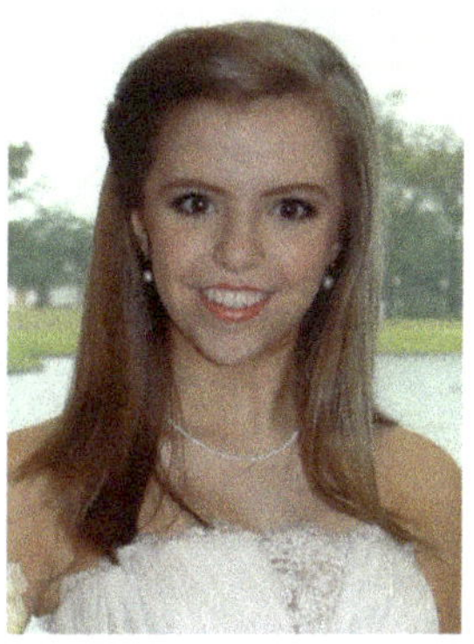

Remi Kimball
BEAUMONT
Daughter of
Thad & Janci Kimball

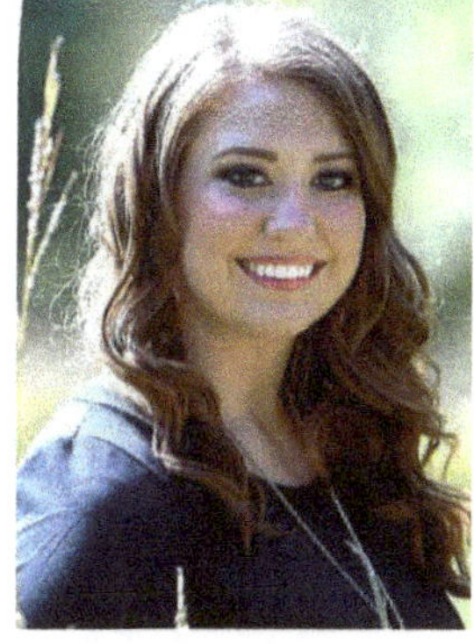

Brooke Loggins
HUNTINGTON
Daughter of
Robert & Alicia Loggins

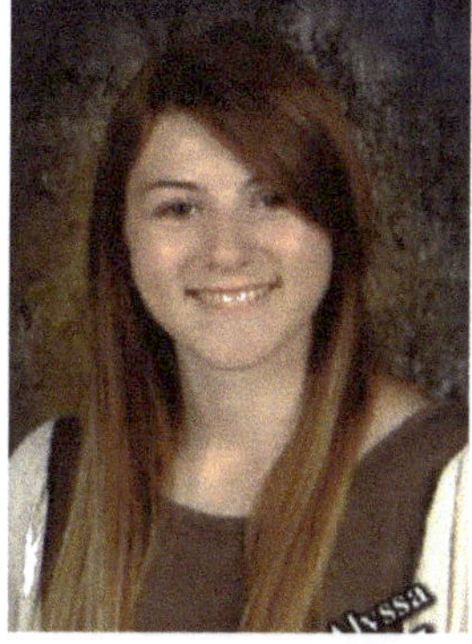

Alyssa Bailey
MONT BELVIEU
Daughter of
Richard & Stacy Bailey

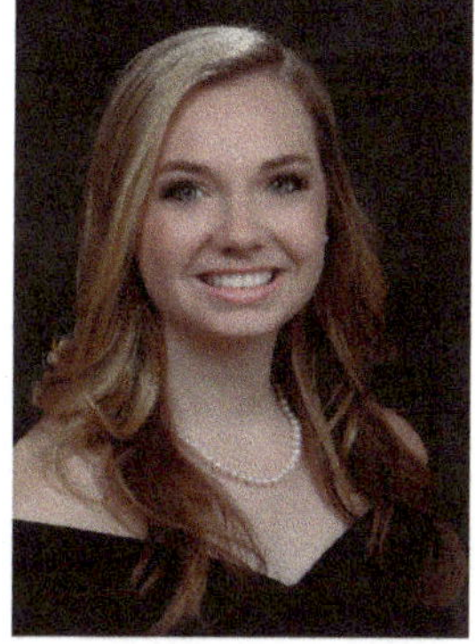

Sarah Ellis
CHINA
Daughter of
Mr. & Mrs. Brian Ellis

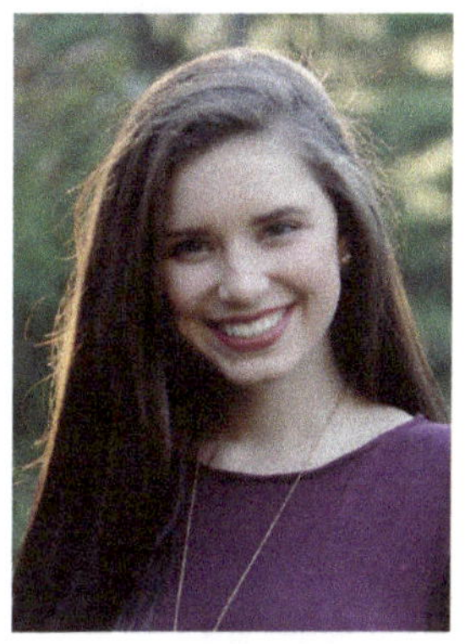

Makensie Cowart
TYLER
Daughter of
Darin &Tammy Cowart

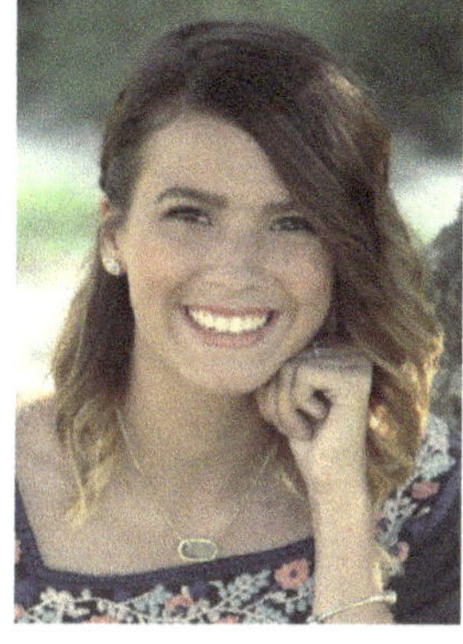

Brooke Wolcott
BROOKELAND
Daughter of
Billy & Kristy Wolcott

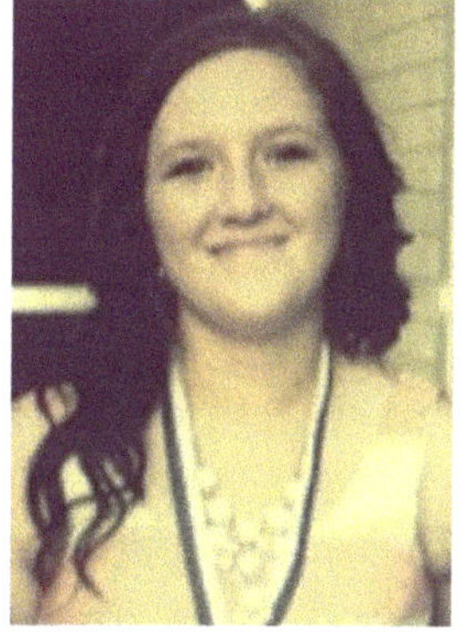

Madison Moody
CORRIGAN
Daughter of
Mason & Misty Moody

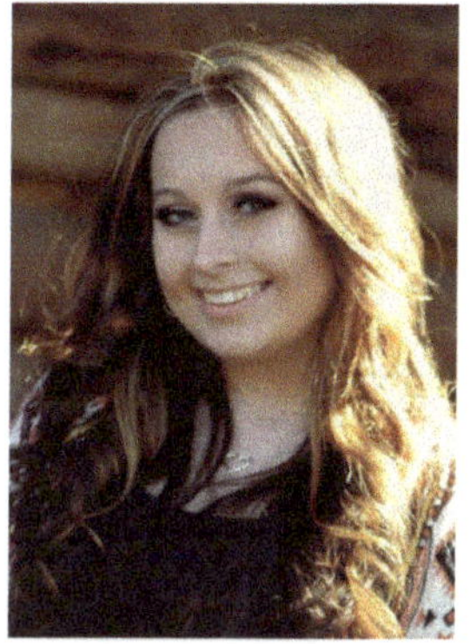

Shania Peavy
SILSBEE
Daughter of
Billy Dan & Stacy Peavy

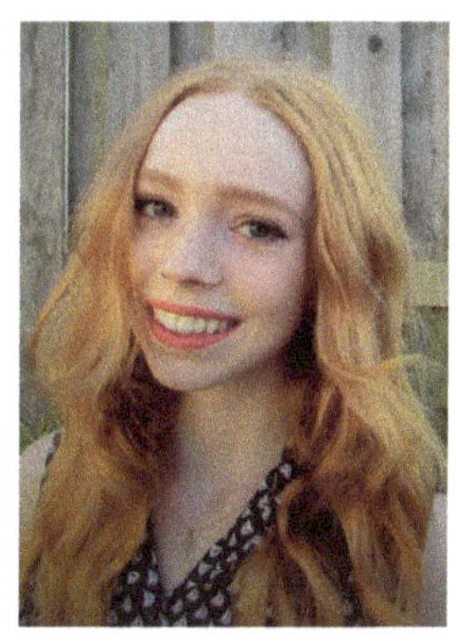

Lillian Joffrion
NEW ORLEANS, LA
Daughter of
Trishell Joffrion & Phillip Joffrion

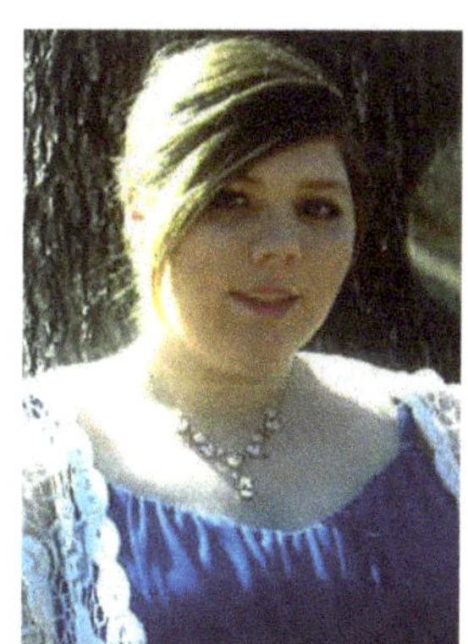

Jessica Bronson
HUNTINGTON
Daughter of
Marshall Bronson & the late Helen Bronson

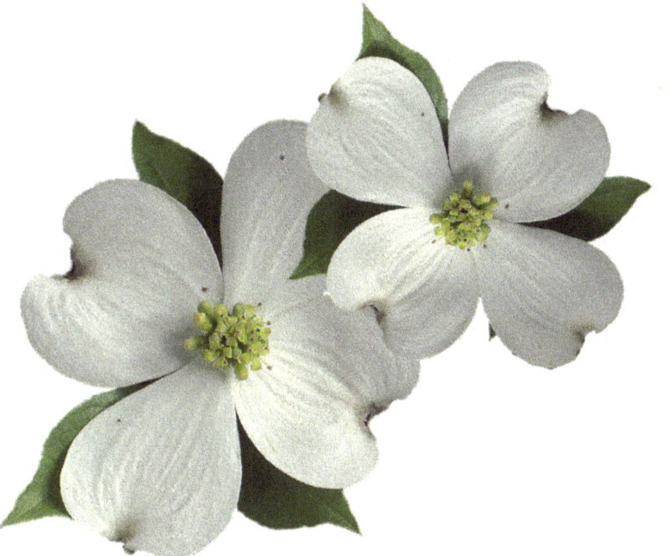

"The Father of the Dogwood Festival"

James Edward Wheat

January 6, 1887 - October 10, 1968

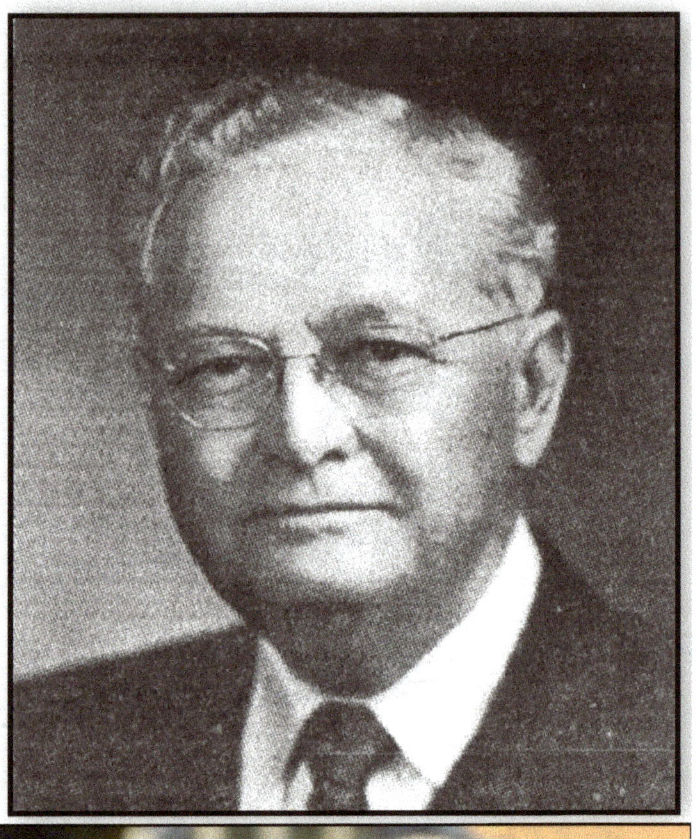

Wheat was born, raised and spent virtually all his life in Tyler County. He was a leader in his community, in school, municipal, church and civic endeavors. In state and national circles he reached the highest levels of success in the legal profession as Chairman of The Democratic Party in Texas, the first Chairman of the forerunner of The Texas Historical Commission, an early reformer of the Texas Prison System, and other varied capacities. He received citations and accolades from all sides, yet he often stated that his greatest satisfaction was the success of the Tyler County Dogwood Festival. He liked to say in the bulletin each year that the Festival indicates "what can be done in small towns where there is a spirit of good will and broadminded appreciation of the cultural and esthetic values of life which enables people to work together for the common good." That was Jim Wheat's "ethic", his "Citizens Creed".

We are fondly reminded of it at this time each year when we realize that once again, "It's Dogwood Time in Tyler County."

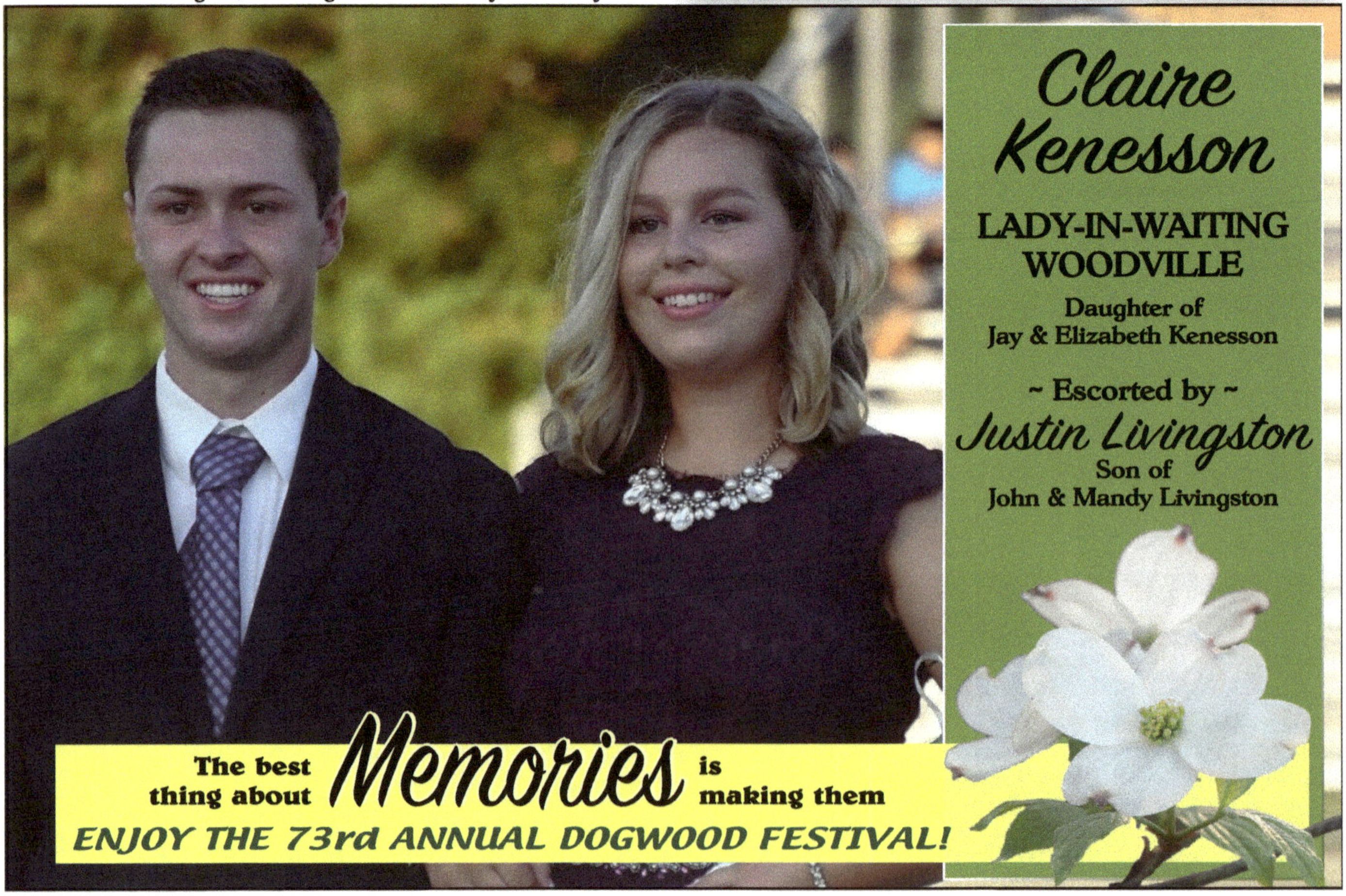

James E. Wheat Award

The spirit of the festival is one of voluntary cooperation which has its fullest expression each year as dozens of committees and individuals work together.....in 1975, the operating directors established the James E. Wheat Award in order to recognize truly exceptional leaders, who through the years have given most unselfishly of their time and talents to the Tyler County Dogwood Festival.

The award, named in honor of the founder of the festival, will be presented annually at the discretion of the directors to the person or persons who are felt to have exhibited the spirit of cooperation which James E. Wheat thought was the key to success in this endeavor.

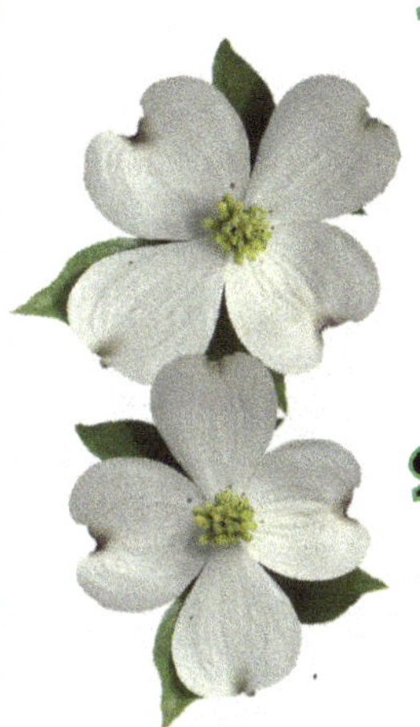

2015 RECEPIENTS

Janay Rainey

Tyler County Heritage Society

Lesha Burkhalter

2016 RECEPIENTS

Jenny Lewis

Tammy Watts

Ted Watts

Previous Recipients of the JAMES E. WHEAT AWARD

F. M. Archer	Webb Ashworth	Jamie Pike	Woodville Lions Club	C. D. Woodrome
Ruby Barclay	R. A. & Katherine Jernigan	Mary Lissie McCombs	Margaret Pope	Glen Conner
Wyatt Bell	Inez Swearingen	Joan Graham	Woodville Volunteer Fire Department	Keith Fuller
Bob Belt	Lawrence Rainey	James A. "Jim" Clark	Kathy Carruth	Michelle Martindale
Joanna Bennett	Kaaren Sullivan	Oleane Bendy	Vivian Davis	Beverly Reese
Sue Bracken	Clinton Currie	Cecil Fortenberry	Blanche Risinger Shaw	Trey & Sharice Allison
Marie Carter	Jane Worthy	Betsy Stafford	Business & Professional Women	Terry & Melissa Riley
Morris Clemmons	Billie Polk	Woman's Reading Club	Judy Brown	Carolyn Collier
J. B. Coffman	George Jarrott	Woman's Study Club	Sassy Scrappers	Steve Evans
Lula Collier	Bevis & Jane Minter	Patricia Brown	Benja McCluskey	Donece Gregory
Maurine Gassiott	Alice McGovney	Dixie Jarrott	Patti Tucker	Chester ISD
Dr. James Jinnette	Blanche Risinger	Don Shaw	Frances Clemmons Kuehn	Colmesneil ISD
Odessa Lowe	Milton Walters	Clyde Taylor, Sr.	Keith Hays	Spurger ISD
Alford Lee Polk	J. T. Stryker	Belinda Allison	Jerry Carruth	Warren ISD
Monroe Prescott	Jim Gainey	Becky Jinnette	Kathy Weaver	Woodville ISD
John Smitley	Annette Smith	Tom McGuire	Sarah & Buddy Crumpler	
F. B. Sullivan	Virginia Baker	Thelma Minyard	Kim Spell	
Quaidie Sullivan	Laytie Lee Clow	Robbie Evans	Lyle Rainey	
Julie Archer	Fred Sullivan	Sharon Fuller	Jacques L. Blanchette	
Josiah Wheat	Ruth Houston	Sarah Holmes	Bob Boykin	
Ruby Wheat	Margarete Crews	Otis Ray Fortenberry	Erline Ingle	
Dottie Wheat	Johnnie Hickman	Bobby Birdwell	Susan Milam	
Ida Mae Stamps	Irma Hickman	Barbara Fortenberry	Eleanor & Grady Holderman	
Hilda Coats	A. J. Cortez	Charles Spurlock	Rhonda Bigby	
Lynn Williams	Katherine Branch	Gil Tubb	Alicia Scoggins	
Esther Wright	Herb Branch	Juanita Brown	Phillis Tubb	
Ellie Hunt	Paul Walker	Tommie Garner		
Tressie McClure	Judy Watts	American Business Women's Assoc.		
Max Graham				

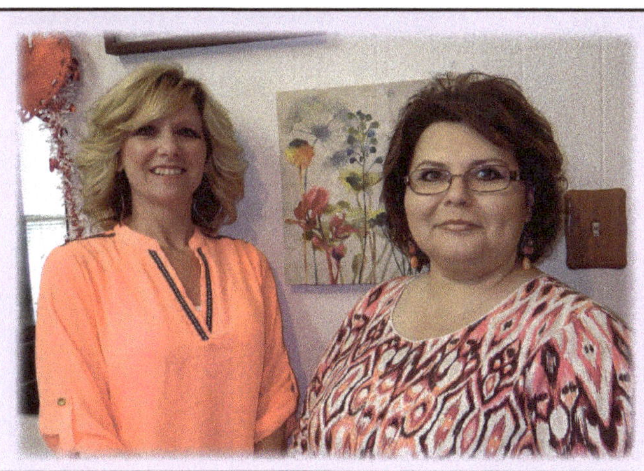

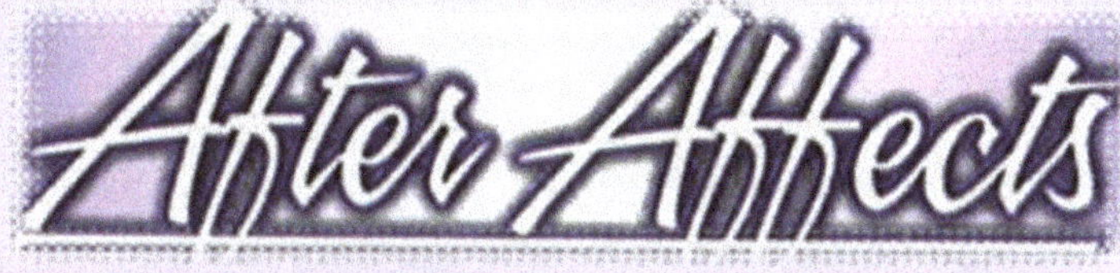

213 CHARLTON STREET • WOODVILLE

409-283-5828

Lesa Simmons ~ Co-Owner, Stylist

Kati Evans ~ Co-Owner, Stylist

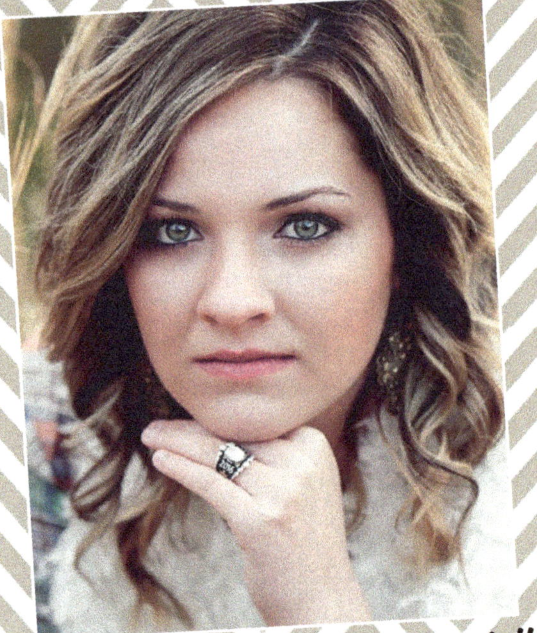

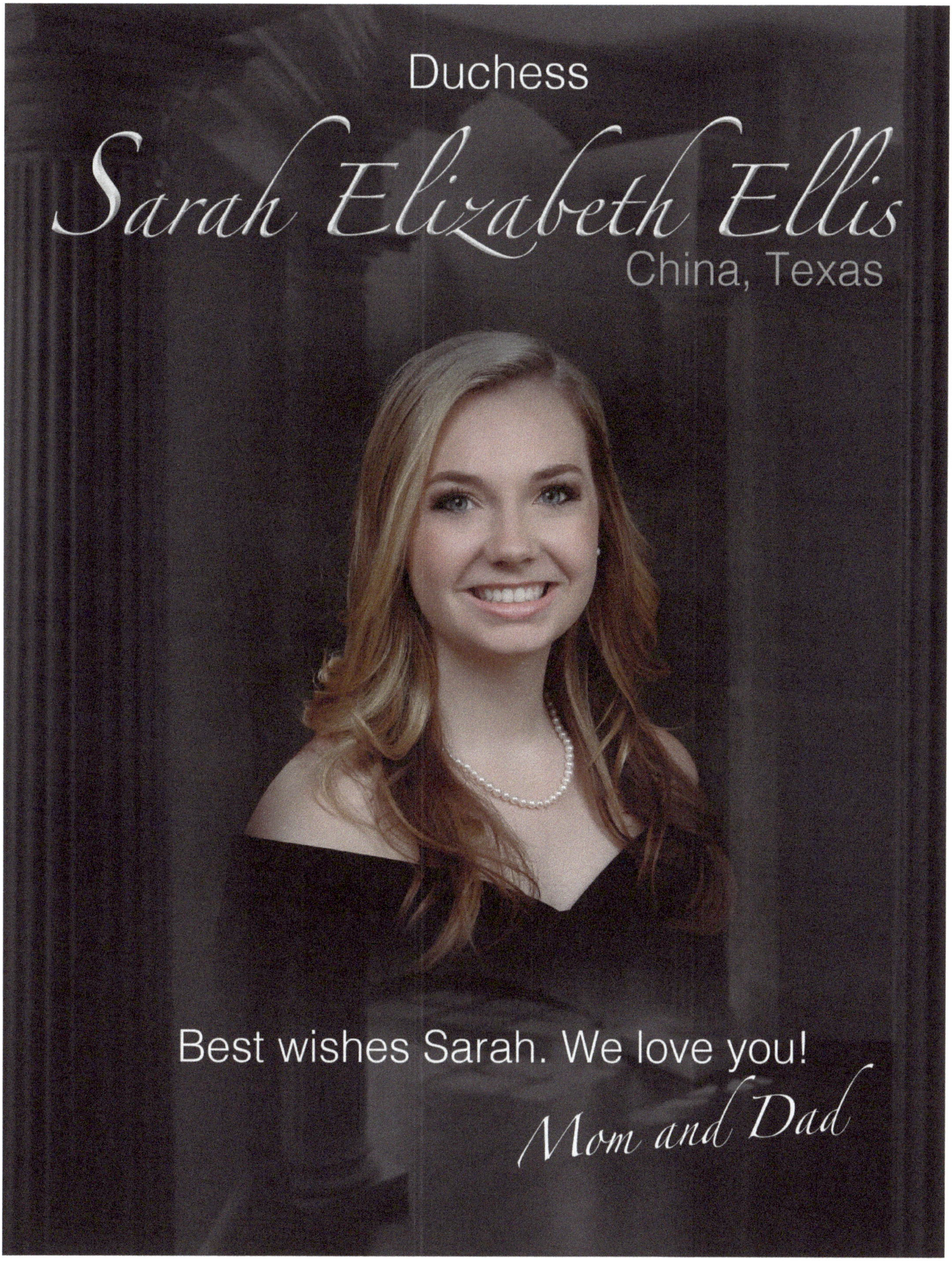

Duchess

Sarah Elizabeth Ellis

China, Texas

Best wishes Sarah. We love you!

Mom and Dad

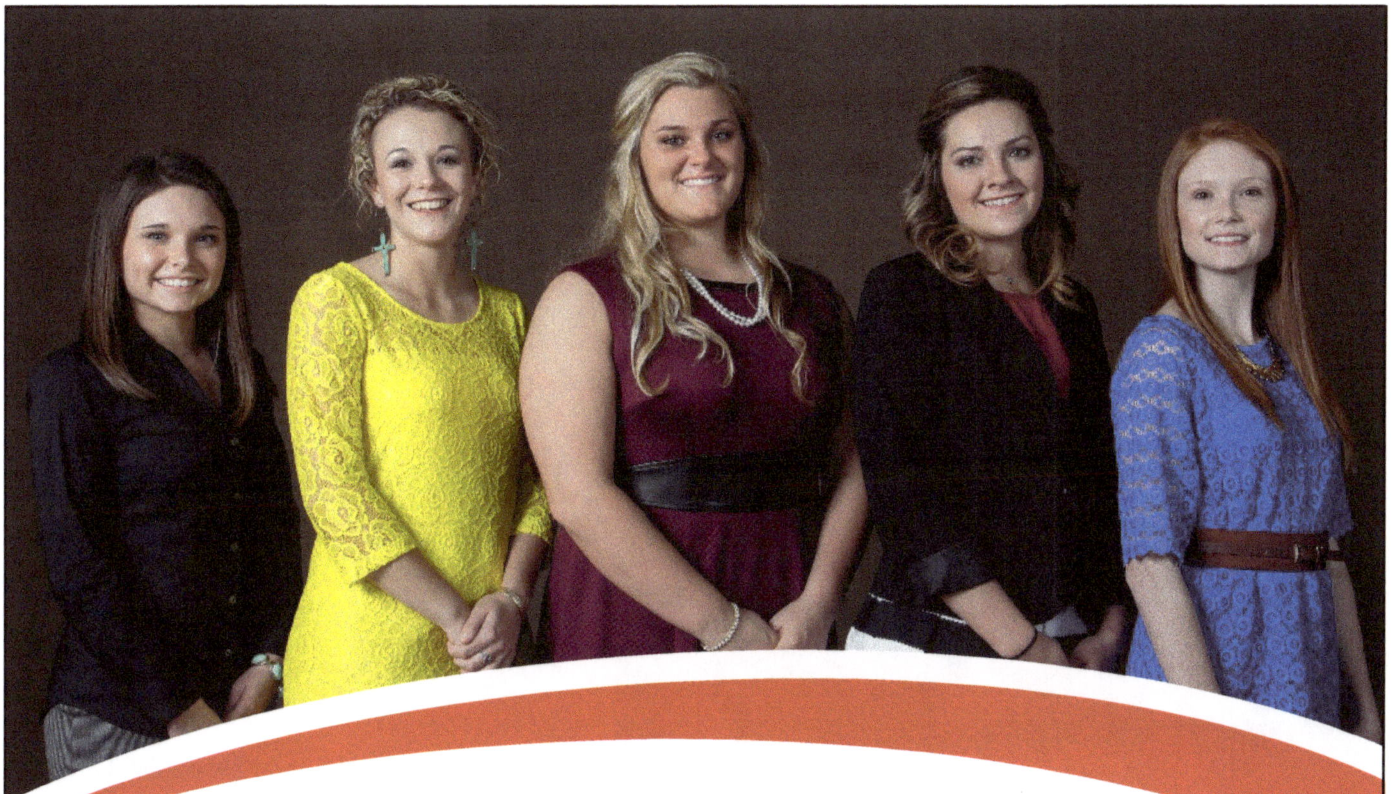

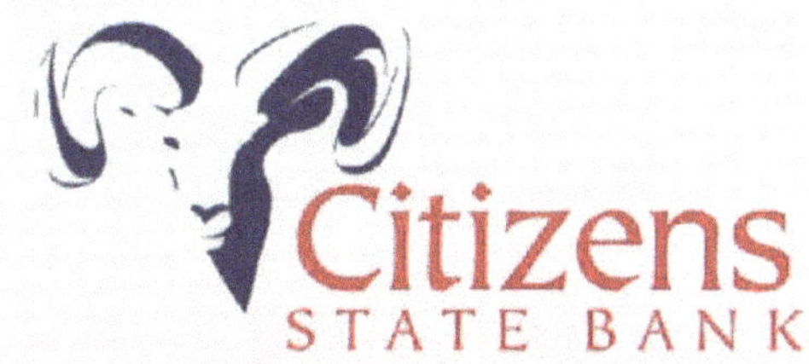

Congratulations to our 2016
Dogwood Princesses

COURTNEY CRAIN OF SPURGER • ASHLYNN WHISNEANT OF WARREN
MIKA MAXWELL OF WOODVILLE • JACI DAVIS OF COLMESNEIL
SHELBY TALLY OF CHESTER

Citizens
STATE BANK

877.343.7348 • CITIZENSBANK.NET
SPURGER • WARREN • WOODVILLE

Member
FDIC

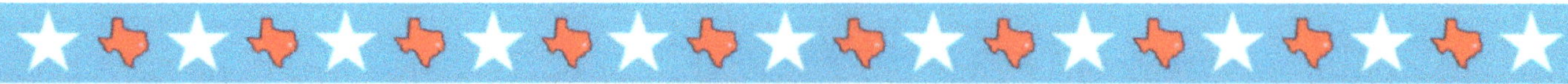

MARTINDALE
Real Estate Investments
To Buy or Sell Call Martindale!

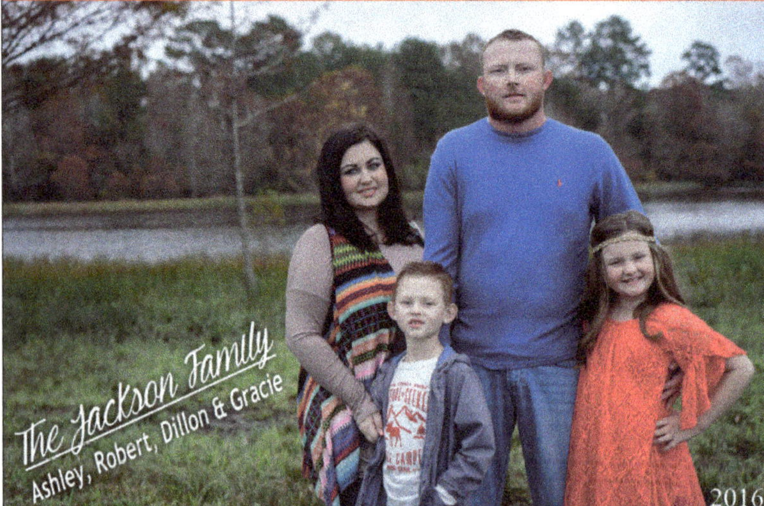

Ashley K. Jackson

ASSOCIATE BROKER, REALTOR®, GRI CERTIFIED AGENT

409-429-4011

www.martindalerealestate.com

Office Addres: 321 S. Magnolia • Woodville, TX 75979

409-283-8727

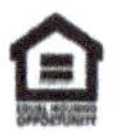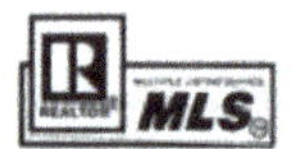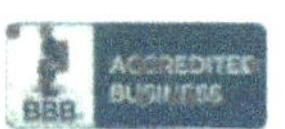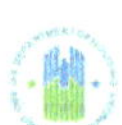

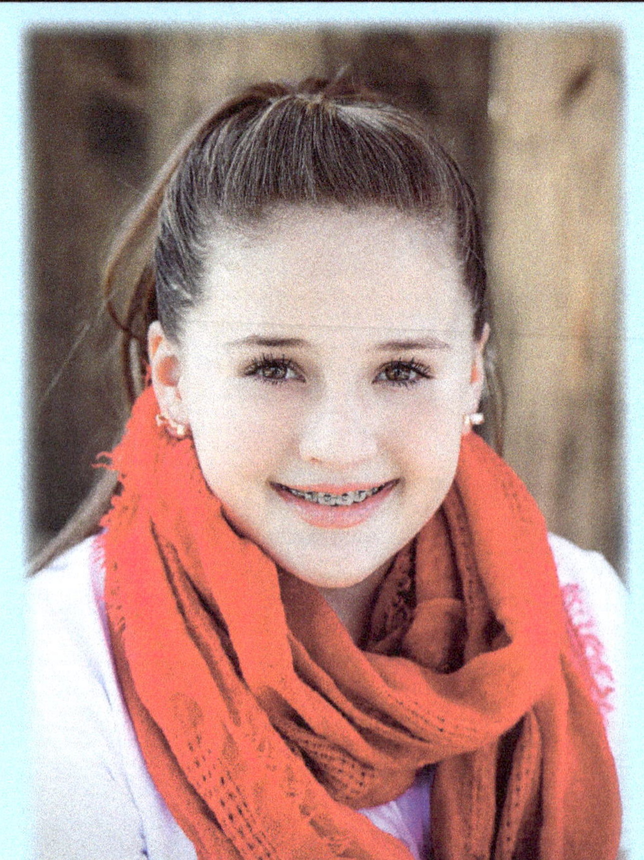

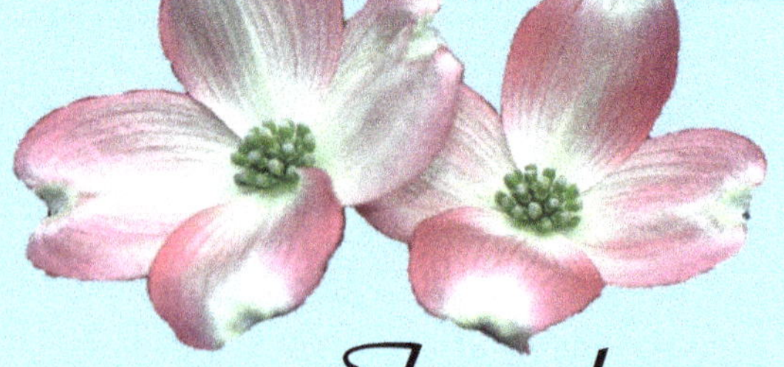

Congratulations

KAELA LEA HASSIG

2016 Trainbearer

WITH LOVE,
Pappy, Gigi, Drew, Mom, Connor, Cash, and McCoy

Dogwood Pharmacy

dogwoodpharmacyrx.com

We can flavor
your medications
with FlavorRx

Flu Shots & Immunizations

LEANING TREE CARDS

205 SOUTH MAGNOLIA • WOODVILLE, TX 75979

409-283-2500

or Rx Phone 409-283-7509

WE'RE OPEN
Monday through Friday
8:30am-5:30pm
Saturday
9:00am-1:00pm

CARSON DEAN CONNER

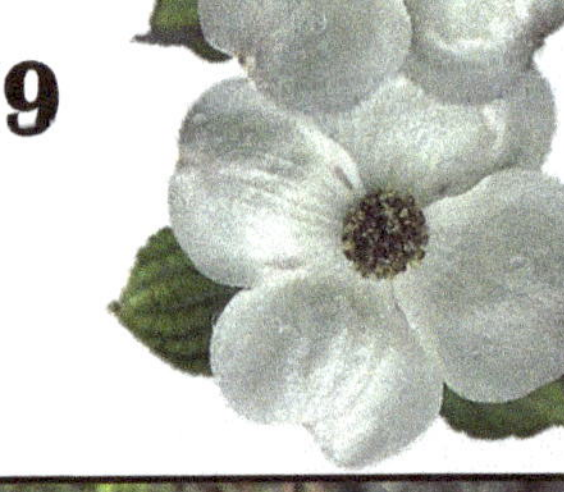

2016
DOGWOOD
FESTIVAL
Scepter Bearer

Carson,

We hope you have a great time at Dogwood! Have fun and make lots of memories!

We are so proud of you!

Love,

Dad, Mom,
Cannon, Kennedy,
Pop, GiGi, and Nonnie

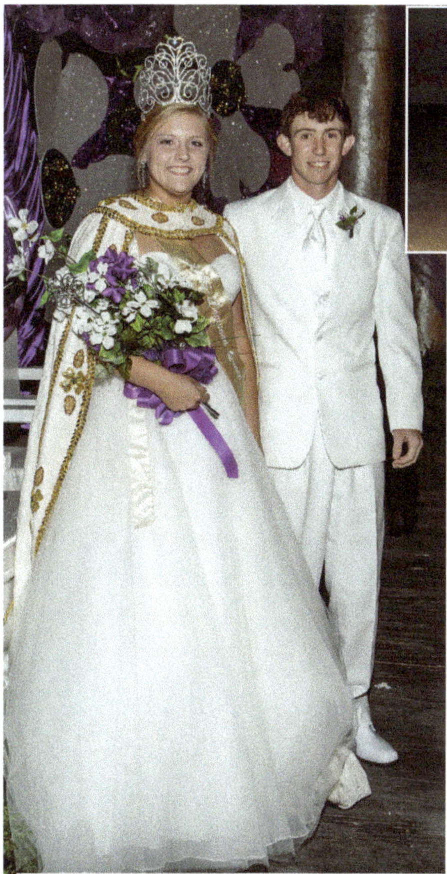

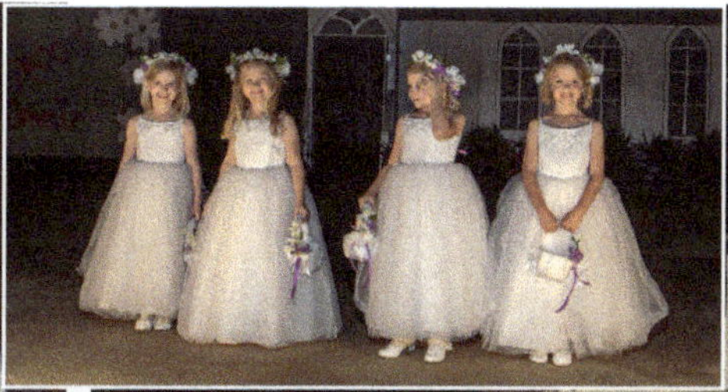

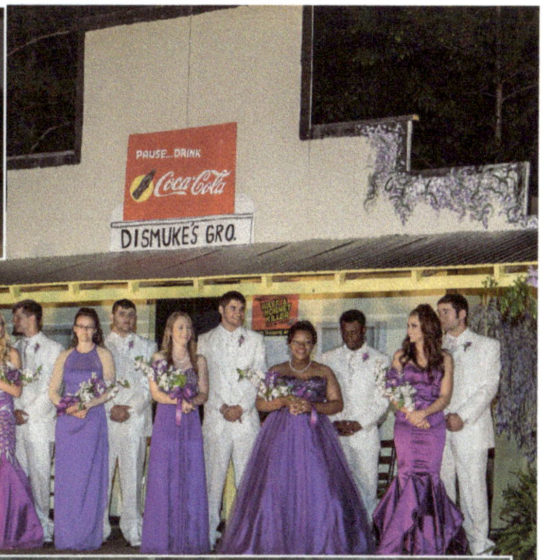

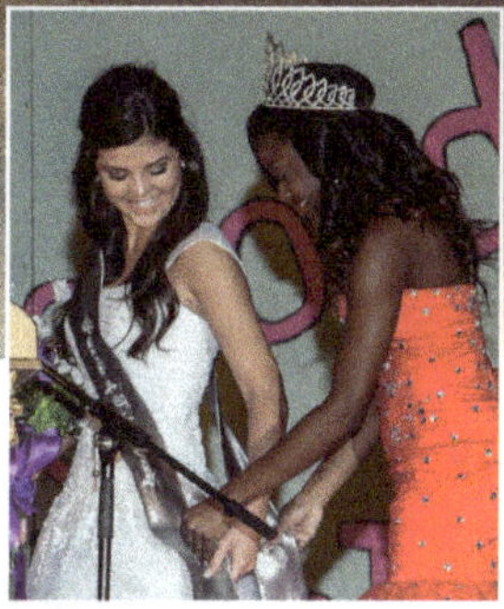

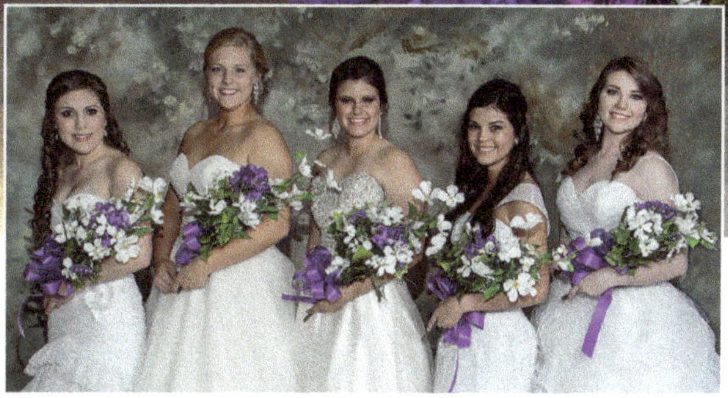

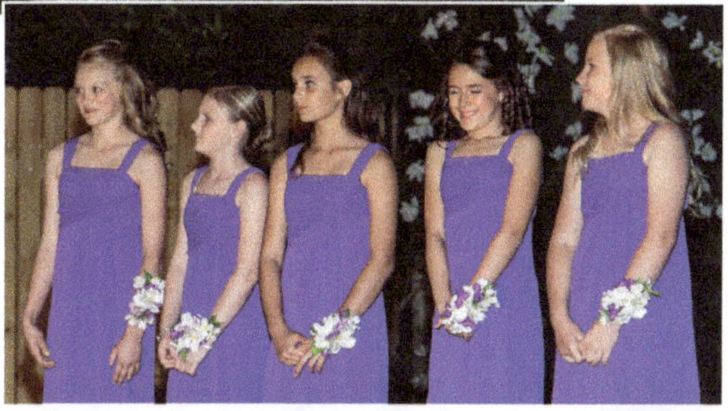

2015 Coronation

DOGWOOD QUEEN
Laken Read

Princess of Tyler County, Rhiannon Odom

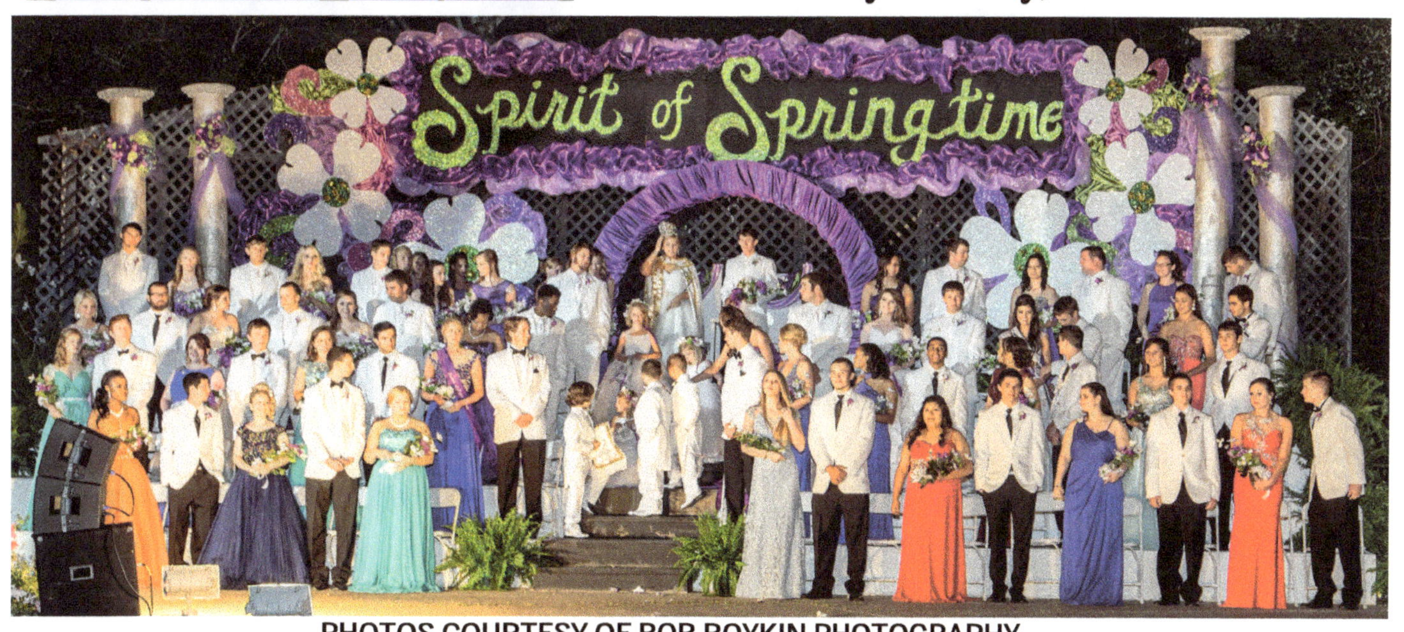

PHOTOS COURTESY OF BOB BOYKIN PHOTOGRAPHY

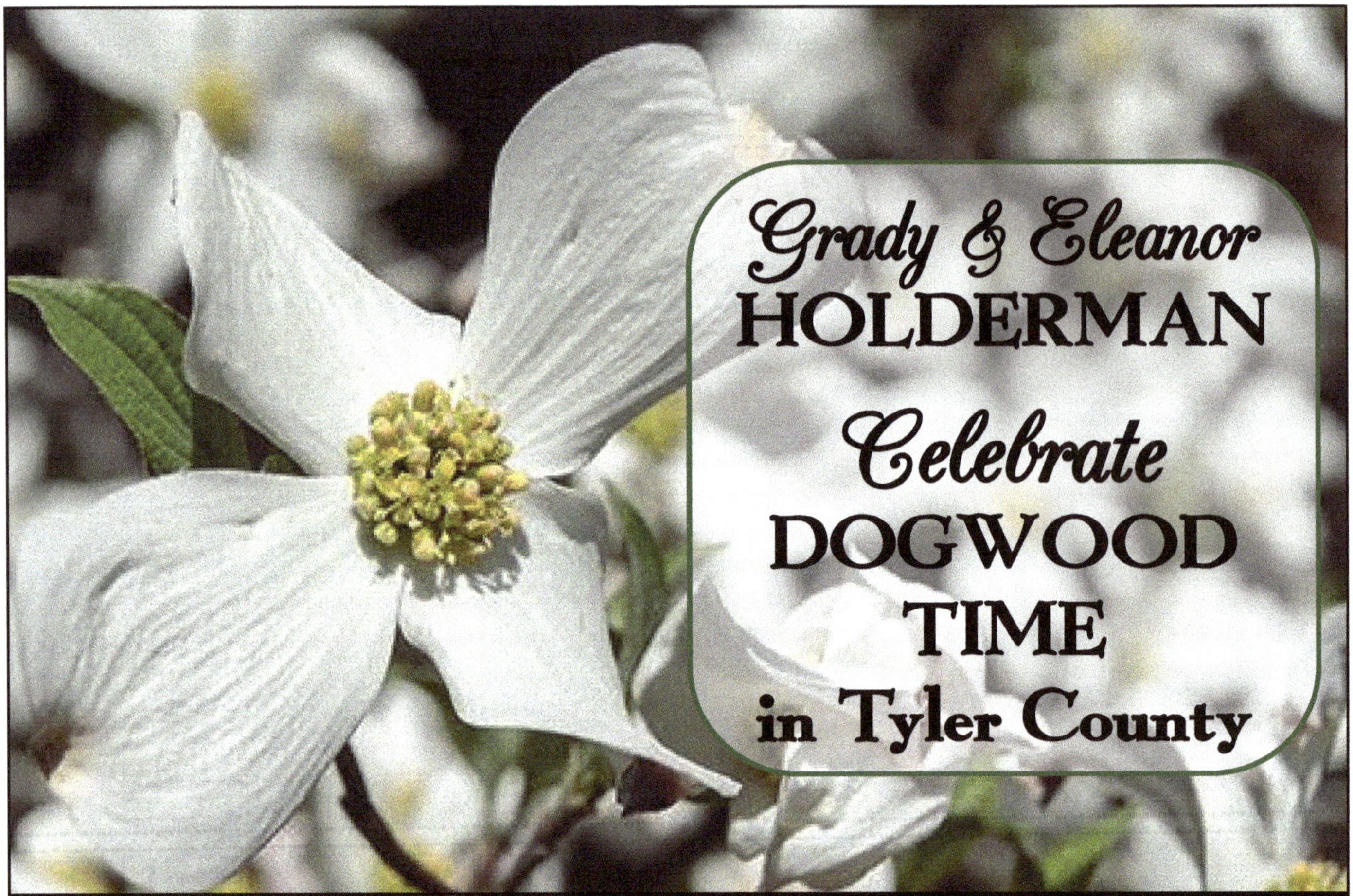

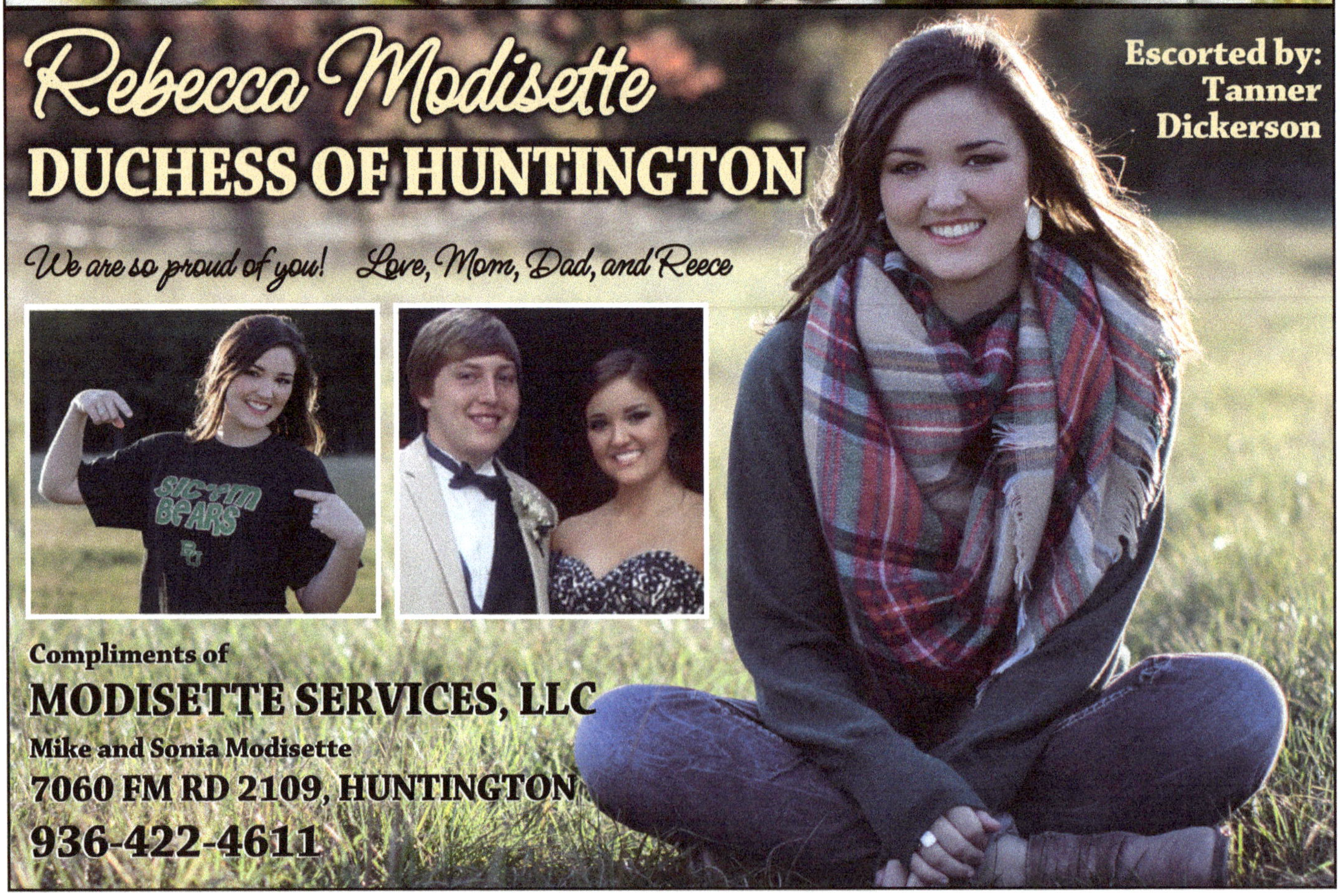

Mika Maxwell
WOODVILLE PRINCESS

We are
so proud
of you!

Love

Mom, John
& McKane

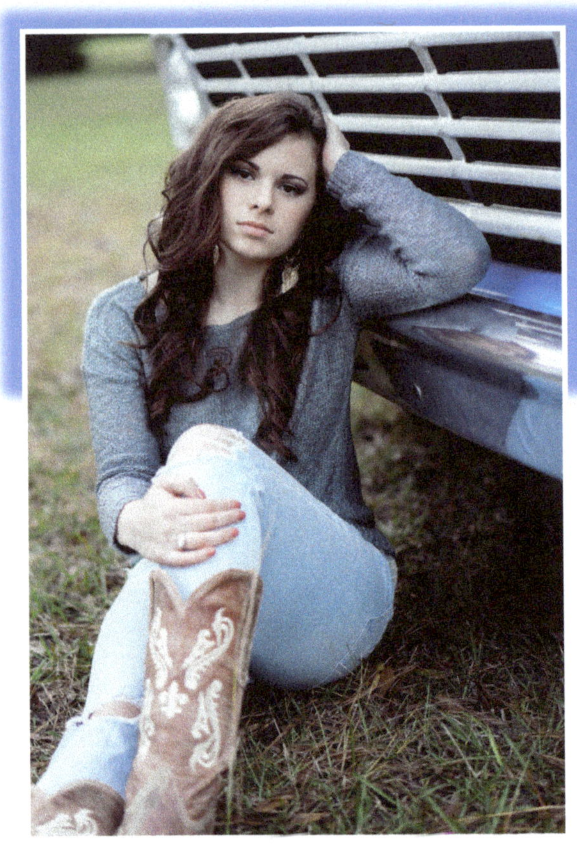

Karley Wood

SPURGER
LADY-IN-WAITING

Daughter of Rickey and Mandy Wood

Granddaughter of
Neil & Kay Means *and* Rickey & Lucille Wood

We Love You!

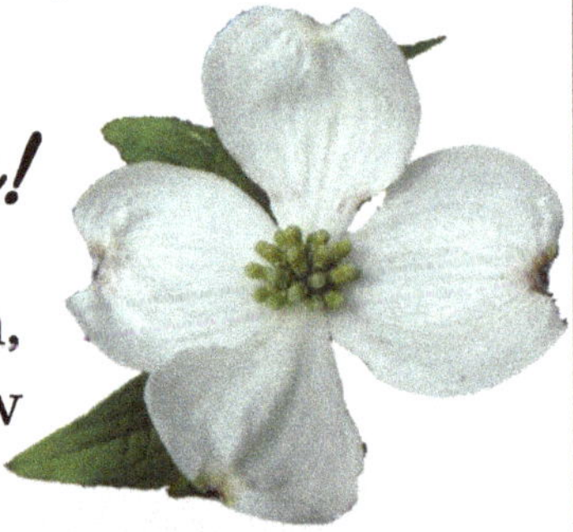

Mom, Dad, Ashleigh,
Nana & Pawpaw

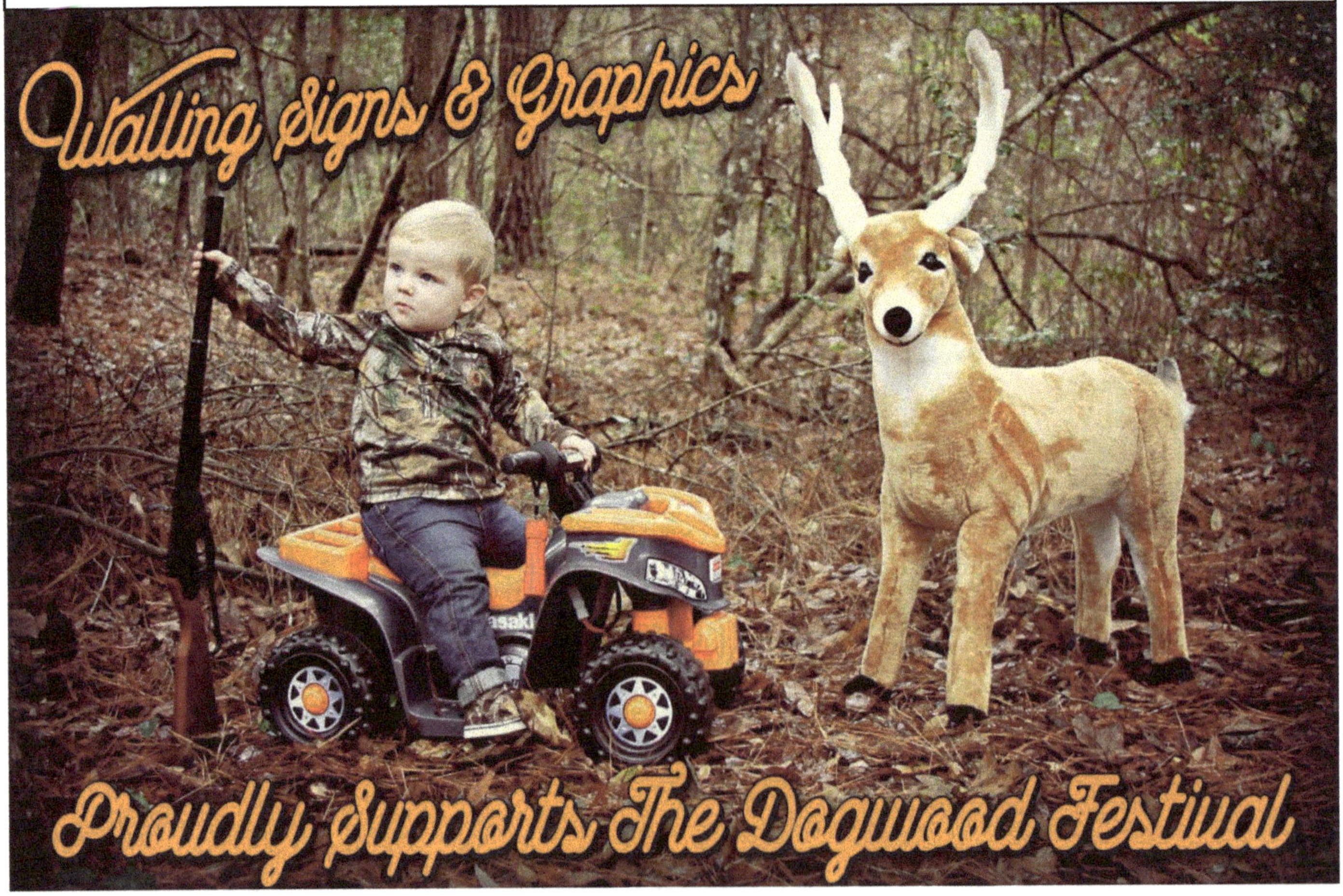

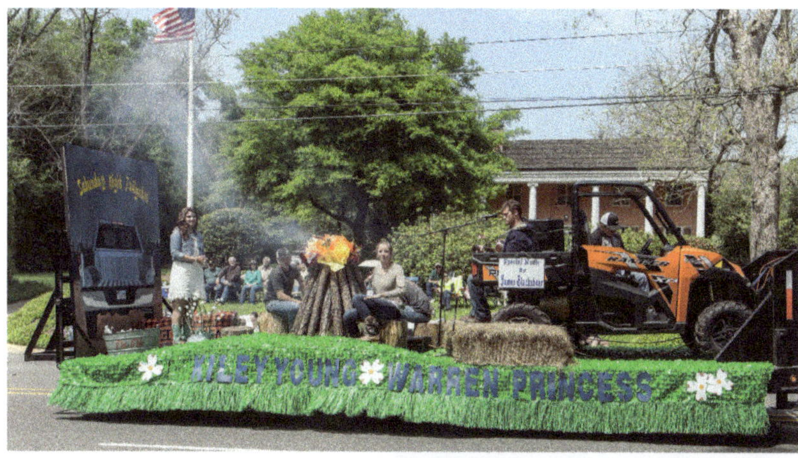

2015 DOGWOOD PARADE

PHOTOS COURTESY OF
BOB BOYKIN
PHOTOGRAPHY

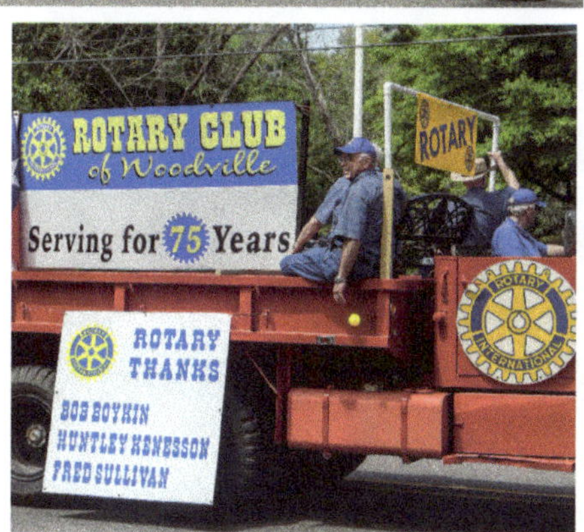

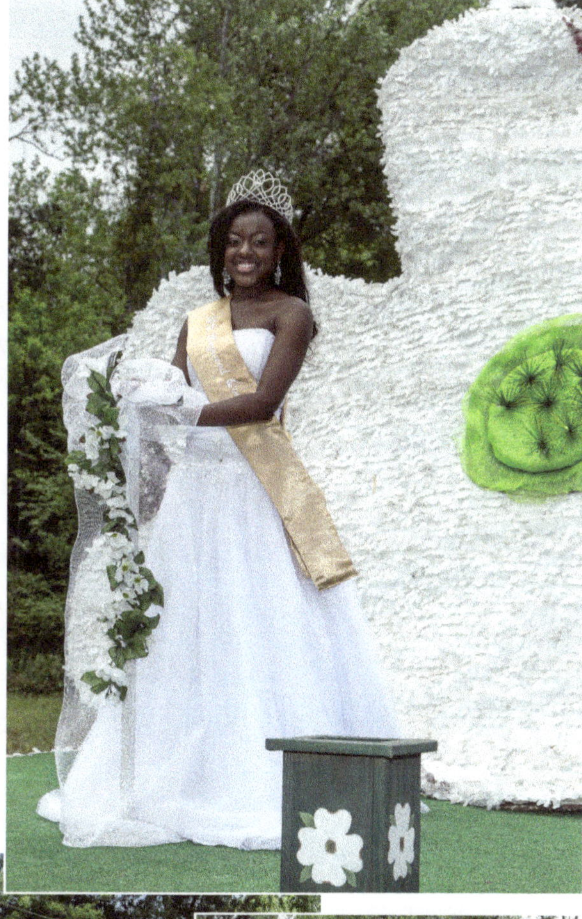

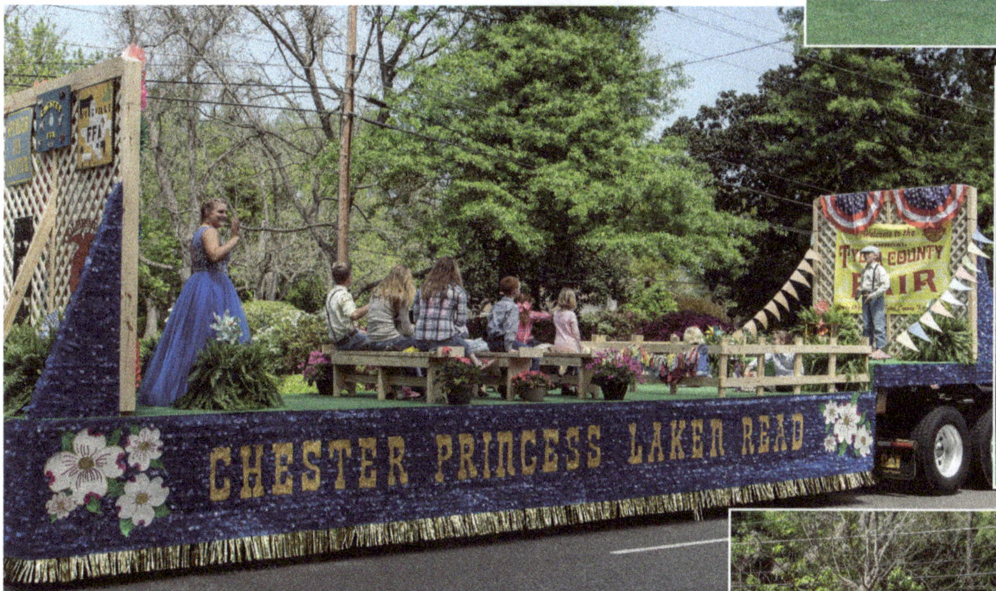

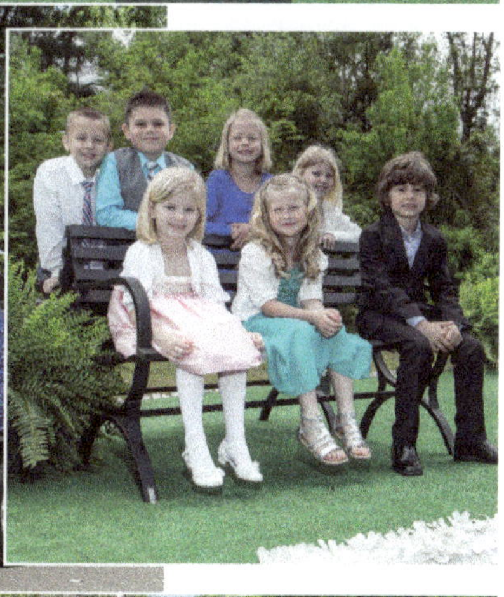

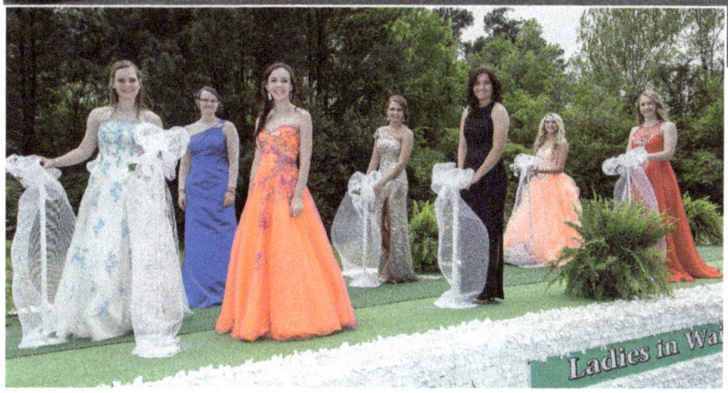

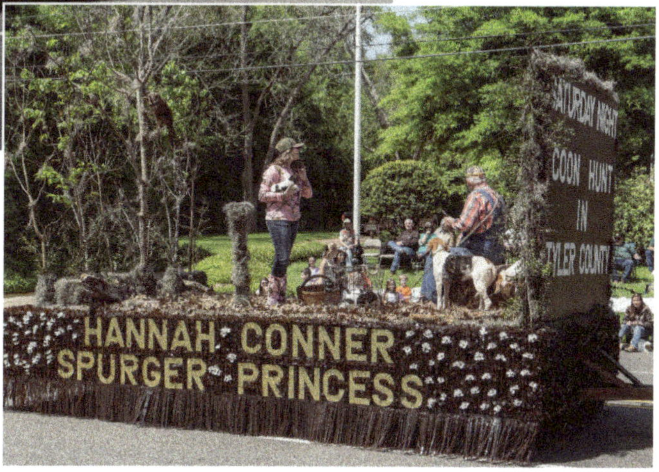

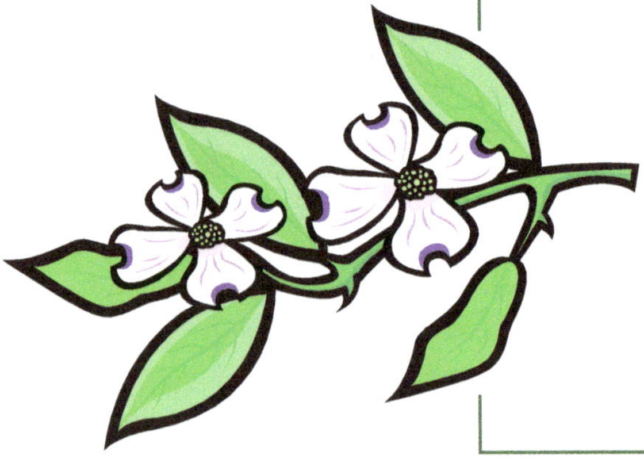

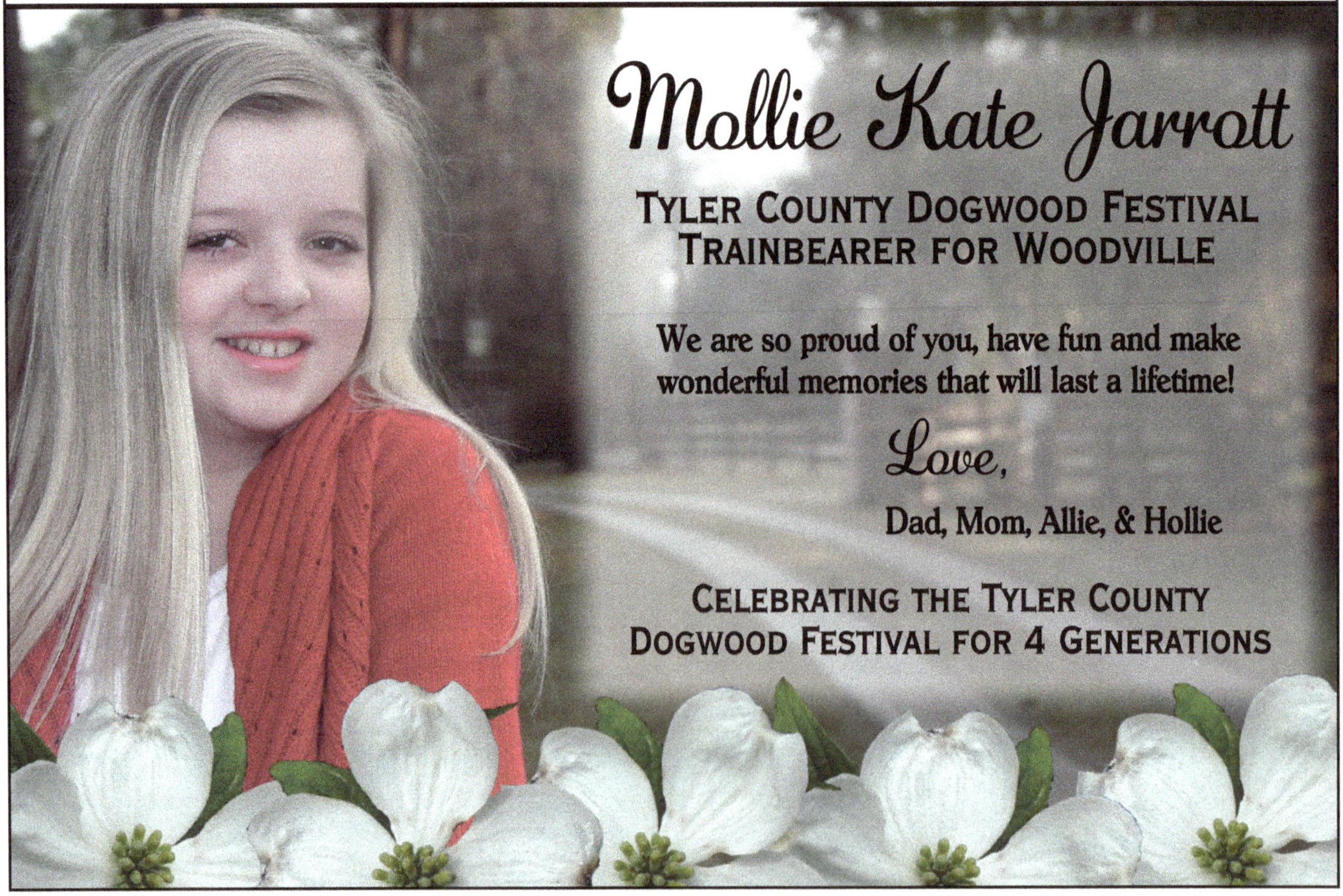

Compliments of

DR. & MRS. JAMES S. JINNETTE

Kingwood, Texas

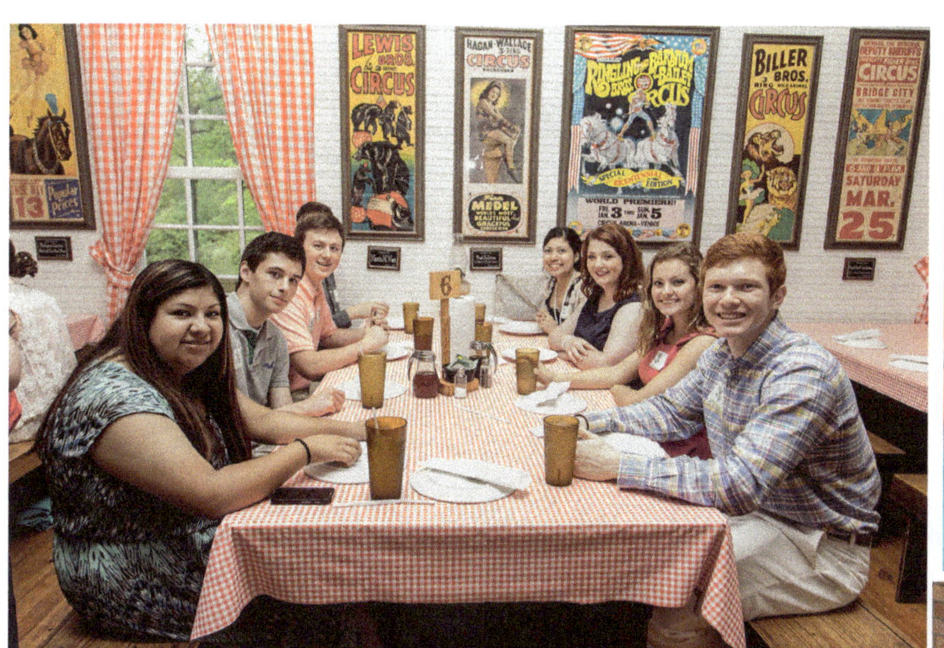

2015 FESTIVAL
Local &
Visiting Royalty

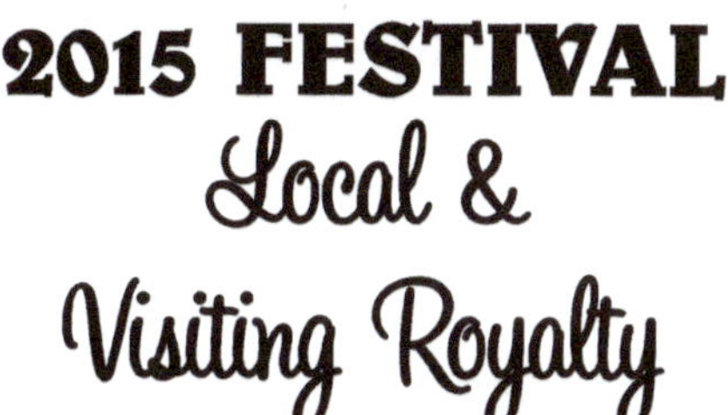

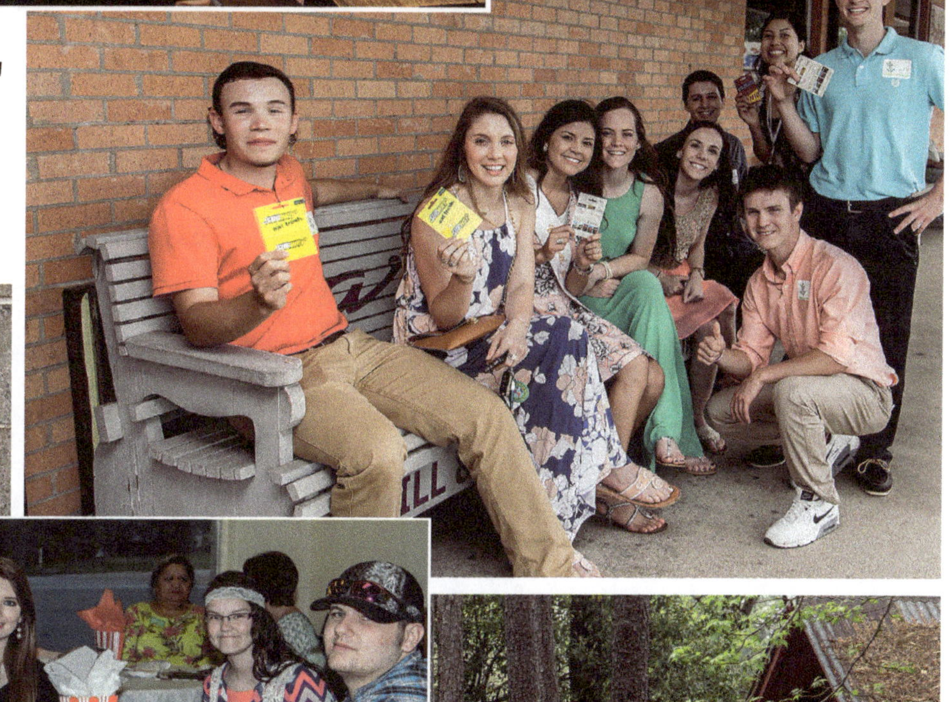

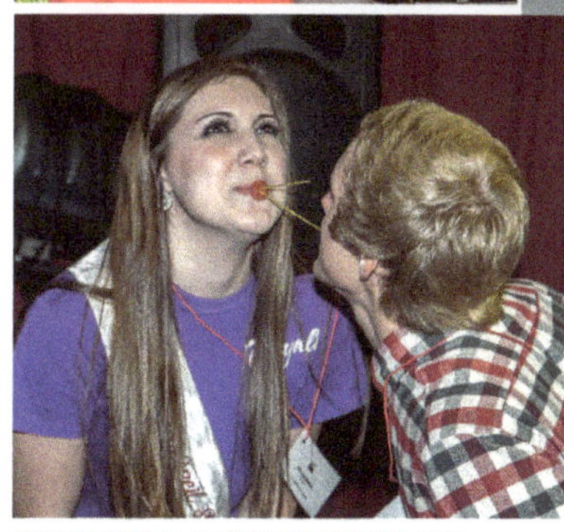

PHOTOS COURTESY OF
BOB BOYKIN PHOTOGRAPHY

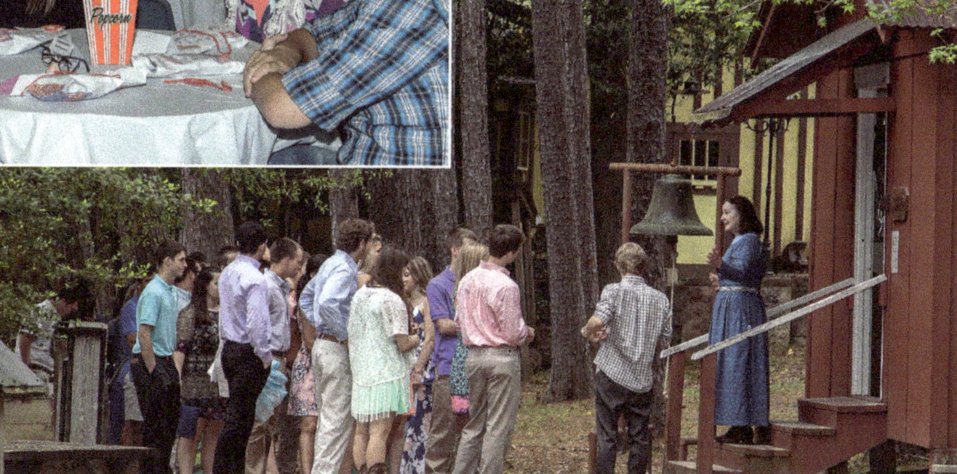

Alyssa Morgan Bailey

DUCHESS OF MONT BELVIEU

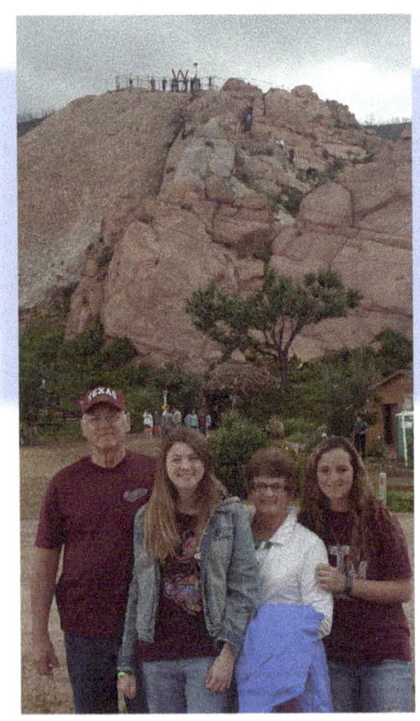

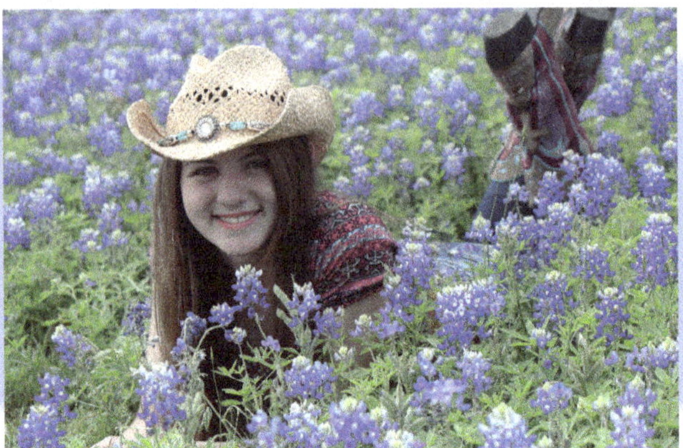

So proud of the young woman you have grown to be!

GRANDAUGHTER OF COACH WILLIE WILSON AND SANDY

DAUGHTER OF RICK AND STACY BAILEY

ALYSSA'S GRANDMOTHER WAS A 1961 LADY-IN-WAITING FROM WARREN

Sandy Heath Wilson

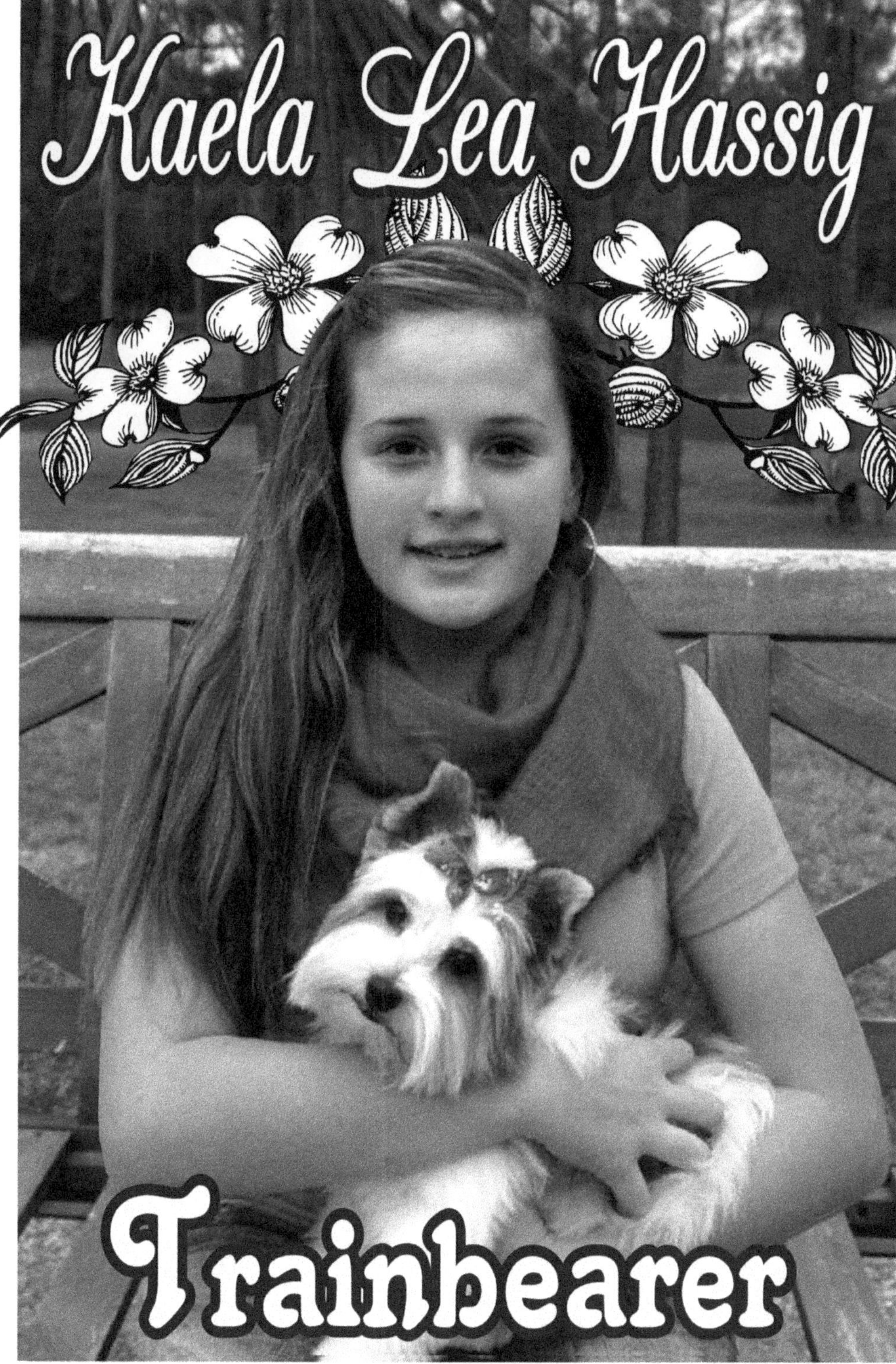

We Love You! Dad, Rosy and Connor

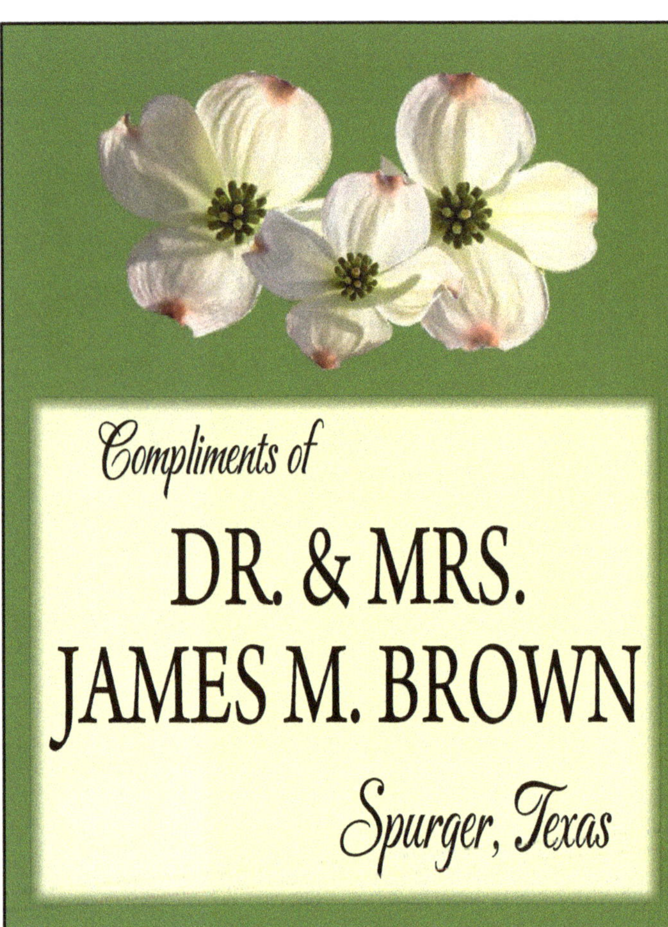

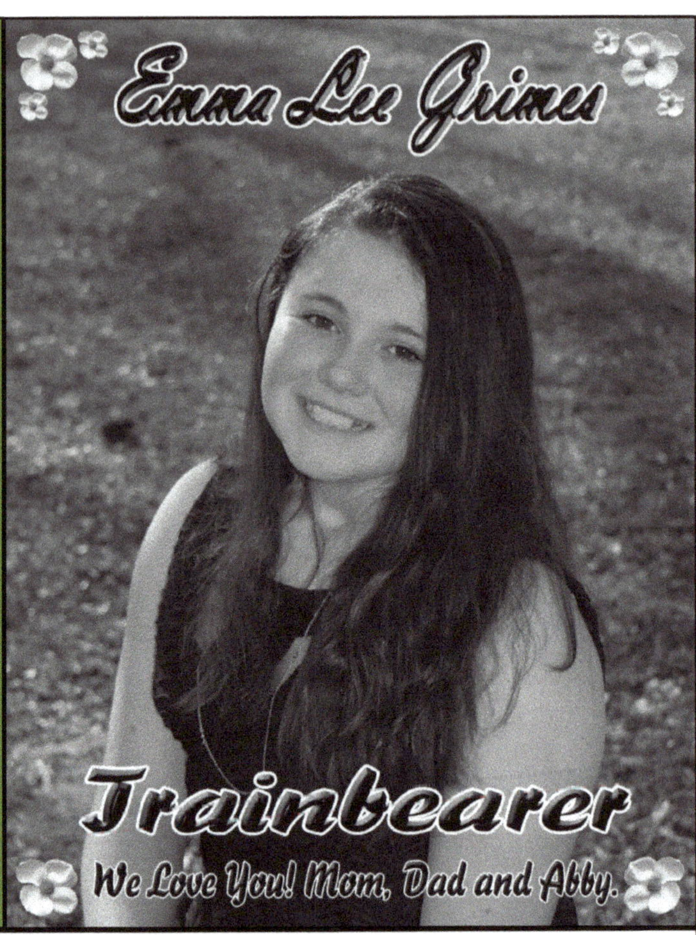

WELCOME TO THE
Tyler County Dogwood Festival!

TRAVIS WEAVER
Agent

2420 US Hwy. 69 S
Woodville, TX 75979

409.283.7101 OFFICE
tweaver@txfb-ins.com

FARM & RANCH / ANNUITIES / MULTI-PERIL / IRAS

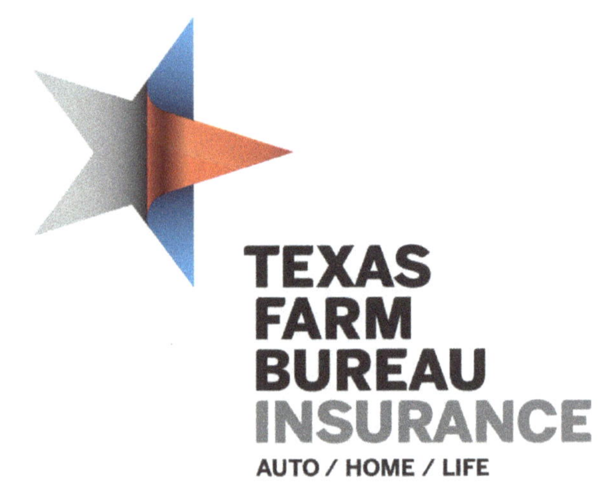

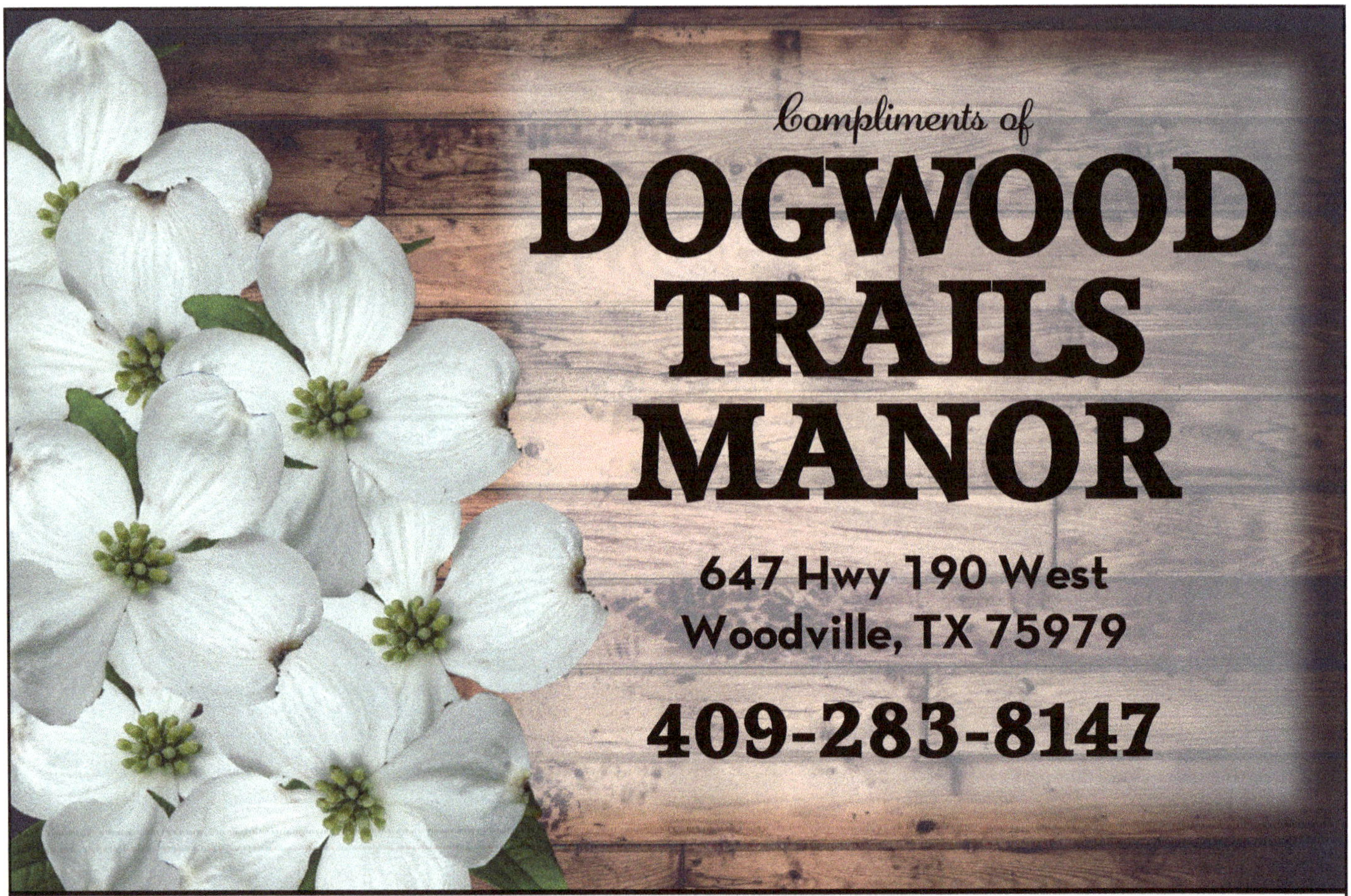

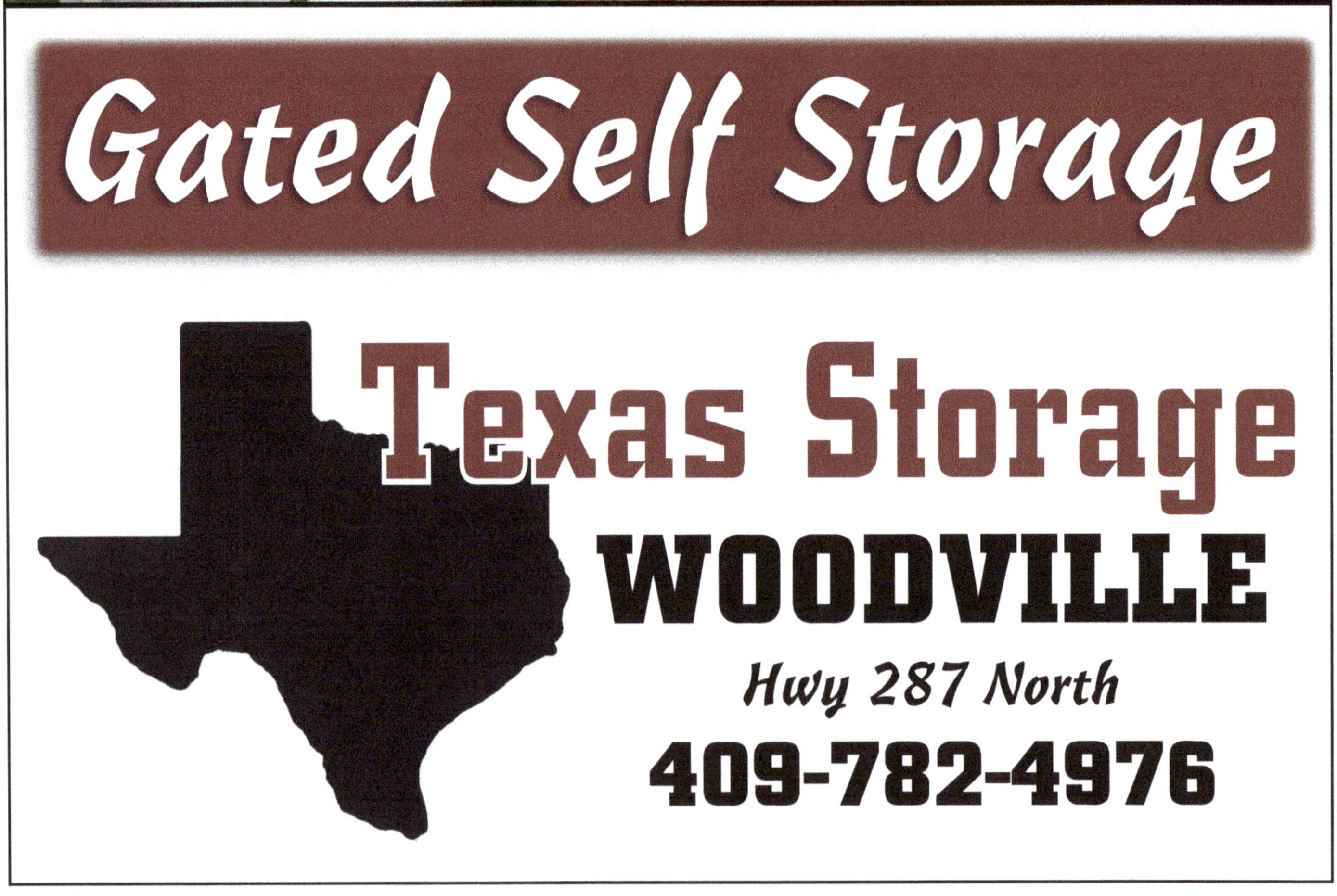

Maddie Kate Read & Conner Read

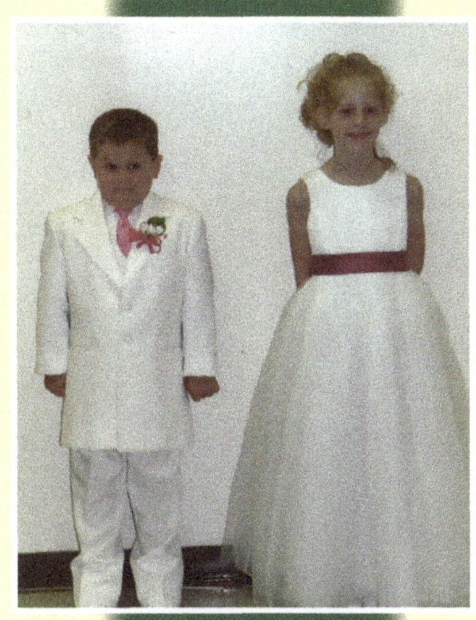

from **CHILDREN'S COURT**
to **QUEEN'S COURT**

We have enjoyed your Dogwood journey.
Make wonderful memories and have
a great time!

Love,
YOUR FAMILY

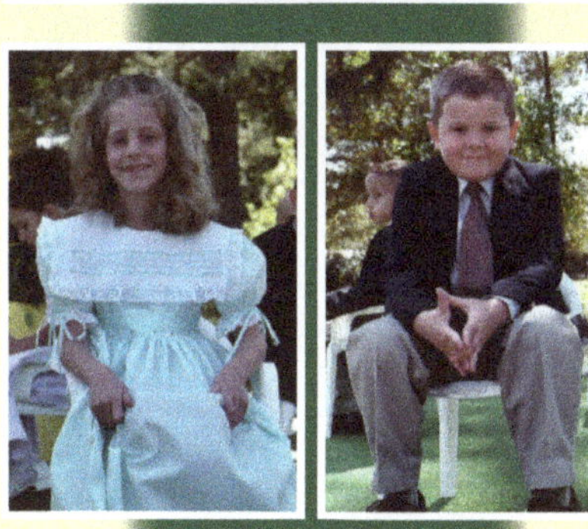

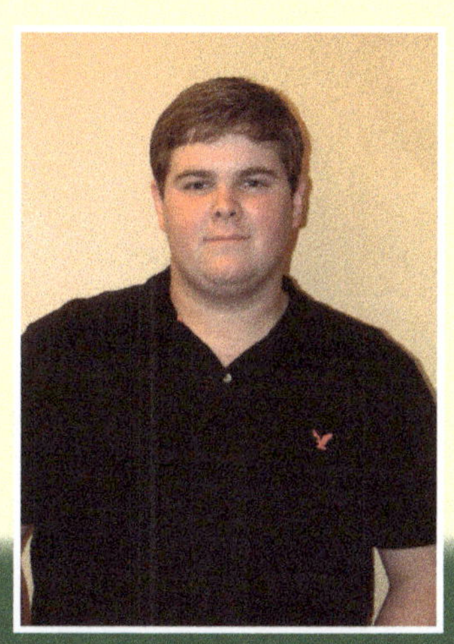

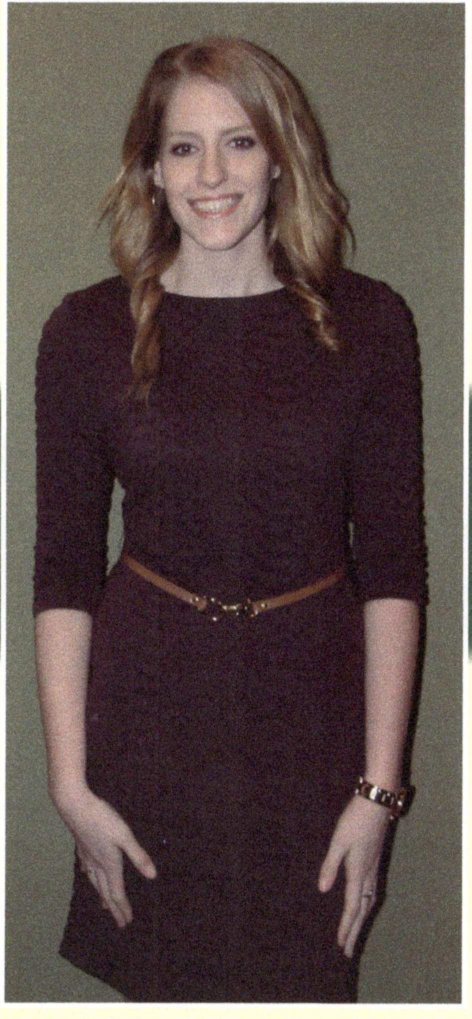

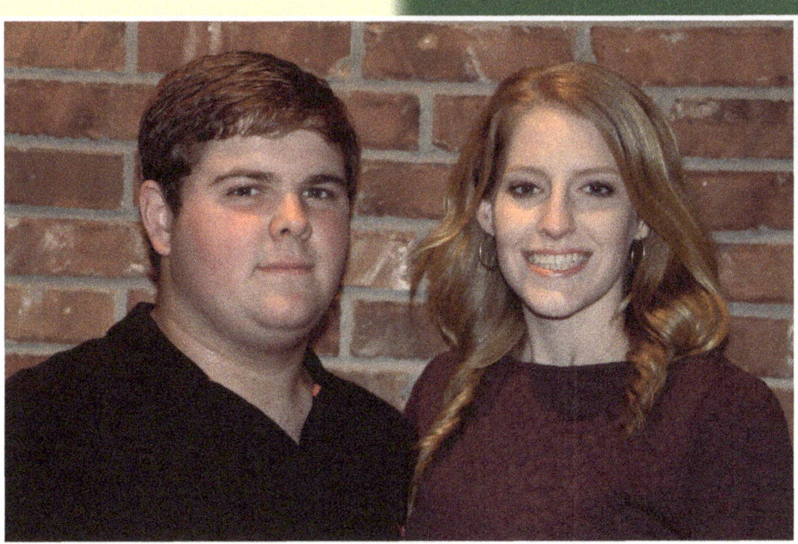

Compliments of Cherie Read

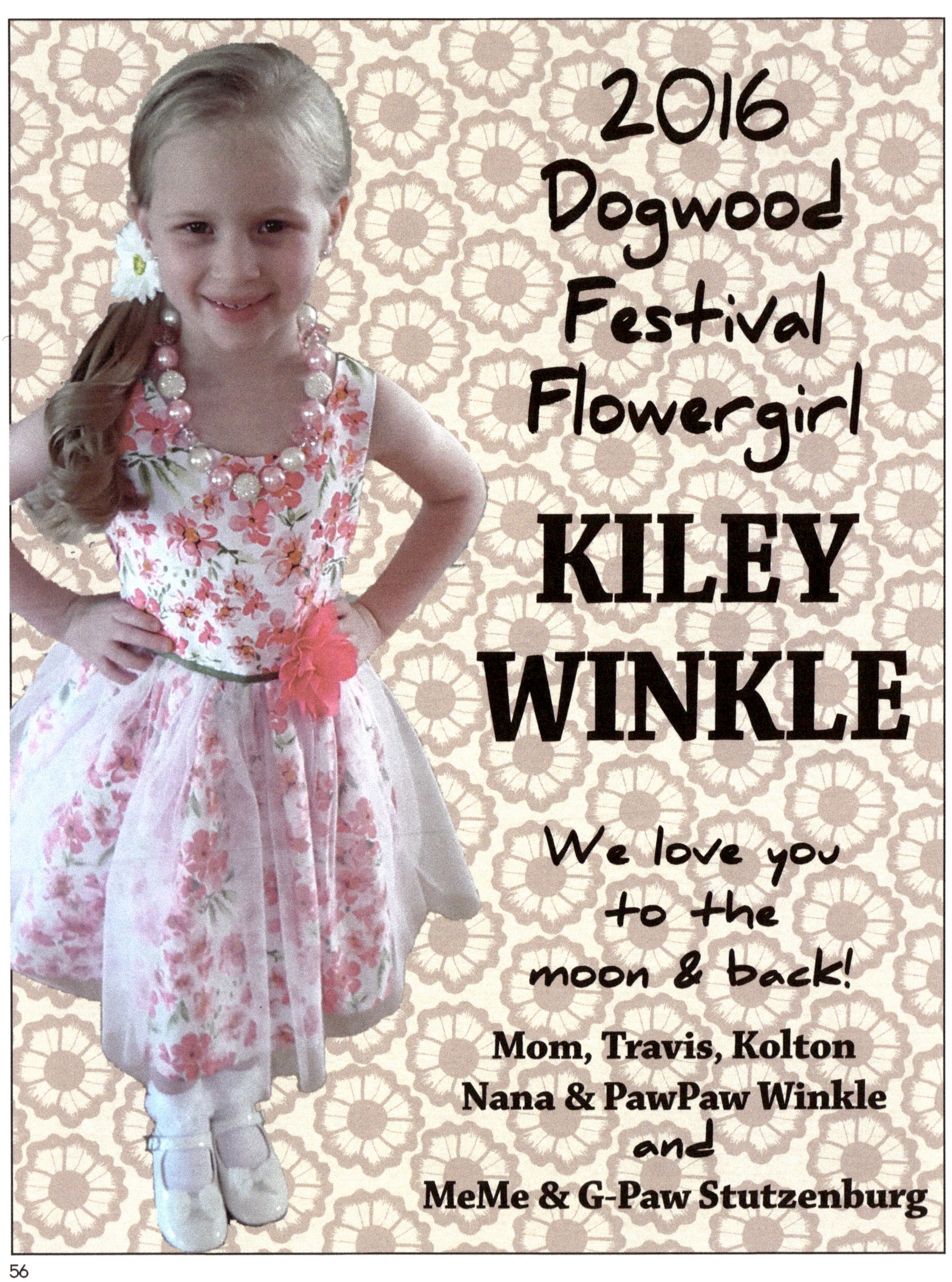

2016
Dogwood
Festival
Flowergirl
KILEY
WINKLE
We love you
to the
moon & back!
Mom, Travis, Kolton
Nana & PawPaw Winkle
and
MeMe & G-Paw Stutzenburg

2015 DOGWOOD PARADE WINNERS

COMMERCIAL FLOATS

1. East Texas Home Health
2. Dogwood Trails Manor
3. Education First CU

NON-COMMERCIAL FLOATS

1. St. Paul's Episcopal School
2. Good Samaritan Fellowship Church
3. Lions Club

PRINCESSES

1. Colmesneil - Morganne Ross
2. Spurger - Hannah Conner
3. Warren - Kiiley Young

DUCHESSES

1. Kelsey Mauer - Kirbyville
2. McKayla Herink - Beaumont
3. Victoria Johnson - Beaumont

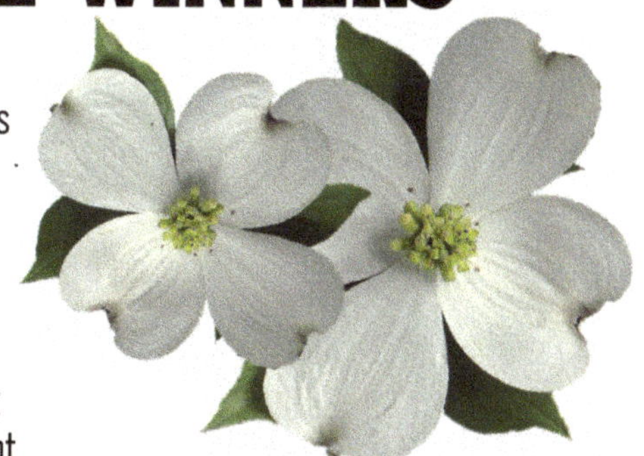

2015 DOGWOOD DASH RESULTS

PHOTOS COURTESY OF BOB BOYKIN PHOTOGRAPHY

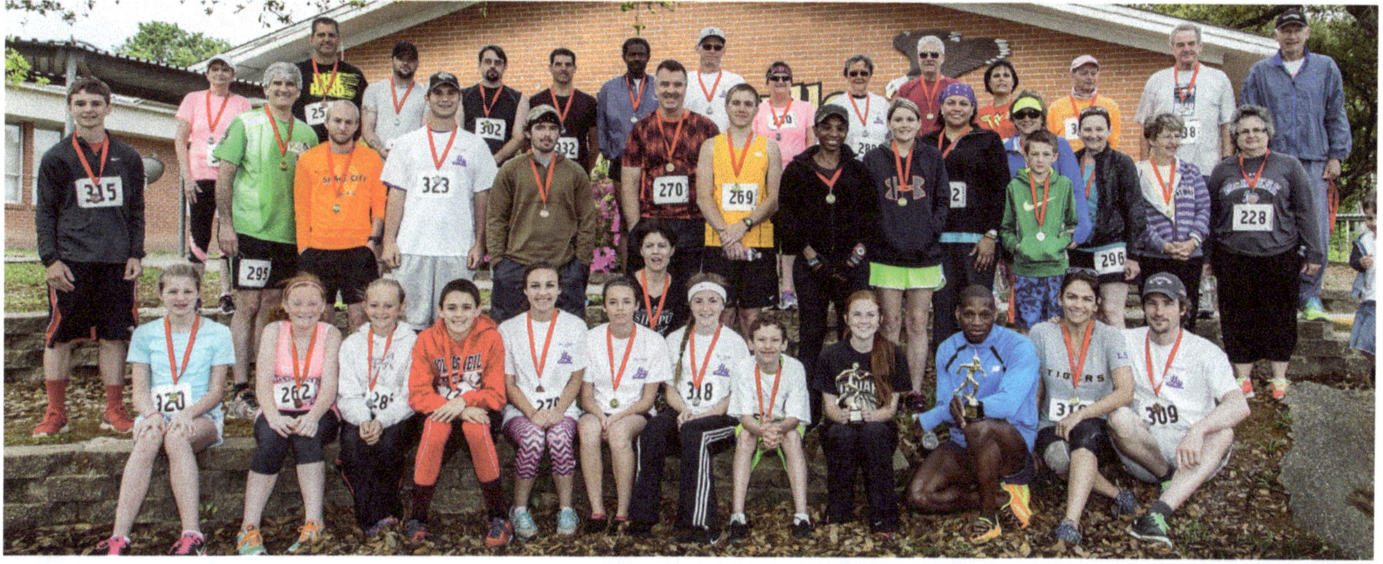

MEN'S OVERALL BEST TIME: Paul Releford 17:26 • WOMEN'S OVERALL BEST TIME: Mallory Walters 21:35

AGE 12 & UNDER	AGE 13-19	AGE 20-29	AGE 30-39	AGE 40-49	AGE 50-59	AGE 60-69	AGE 70-99
MALE:	**MALE:**	**MALE:**	**MALE:**	**MALE:**	**MALE:**	**MALE:**	**MALE:**
Kaleb Powers 21:51	Bobby Swien 19:15	Ryan Kelly 17:45	Bradley Heim 26:26	Armando Bocanegra 18:59	Darren Martin 23:18	Tony Taylor 23:34	Ken Hodges 30:32
Rex McGehee 23:36	Colten Martin 19:49	Joseph Boyd 19:28	Billy Costlow 26:50	Robert Lebouef 20:15	Charles Butler 24:19	Wilford Smith 29:27	Ken Johnson 36:02
Garrett Darder 26:02	Kameron Lindsey 22:12	Sampson Graham 20:16	Lucas Babin 30:24	Jason Drake 22:53	Bobby Swien 34:54		Gerry Simpson 37:17
FEMALE	**FEMALE**	**FEMALE**	**FEMALE**	**FEMALE**	**FEMALE**	**FEMALE**	**FEMALE**
Melody Sanford 25:17	Jackie Hughes 22:30.2	Autumn Hillhouse 24:31	Crystal Valderez 25:35	Luciana Babin 28:24	Gloria Kamil 25:34	Laurie Martin 29:26	Rita Powell 52:36
Shelby Patrick 25:46	Mia Spradlin 22:30.7	Micah Bumstead 26:21	Jeana Dickins 27:14	Kristi Hughes 28:27	Joyce Boudreaux 32:34	Sharon Shaeffer 41:13	Jean Bright 54:23
Chloe Brown 26:28	Rylee Smith 23:03	Amber Rainey 26:30	Arin Dees 27:27	Julie McGehee 33:04	Catherine Martin 39:16	Diana Kersten 75:40	

WOODVILLE VETERINARY CLINIC

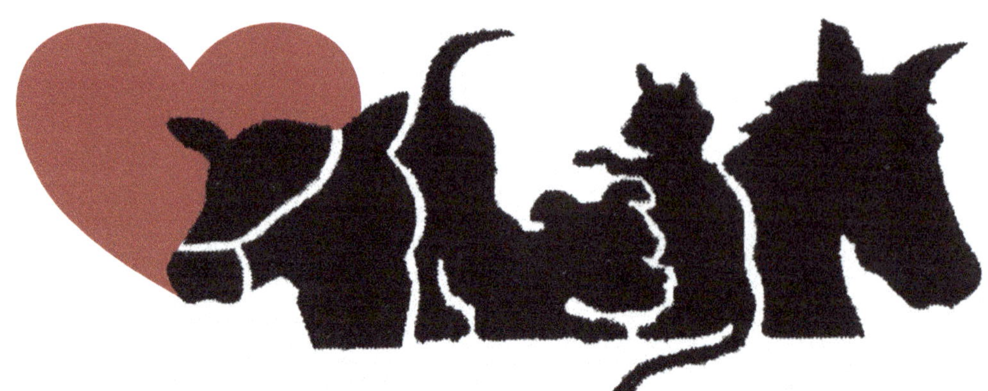

G. P. McCluskey D.V.M.   **409-283-2220**

1351 North Magnolia • PO Box 222

WOODVILLE, TEXAS 75979

Harrison Dillard • 2016 Dogwood Festival Crown Bearer

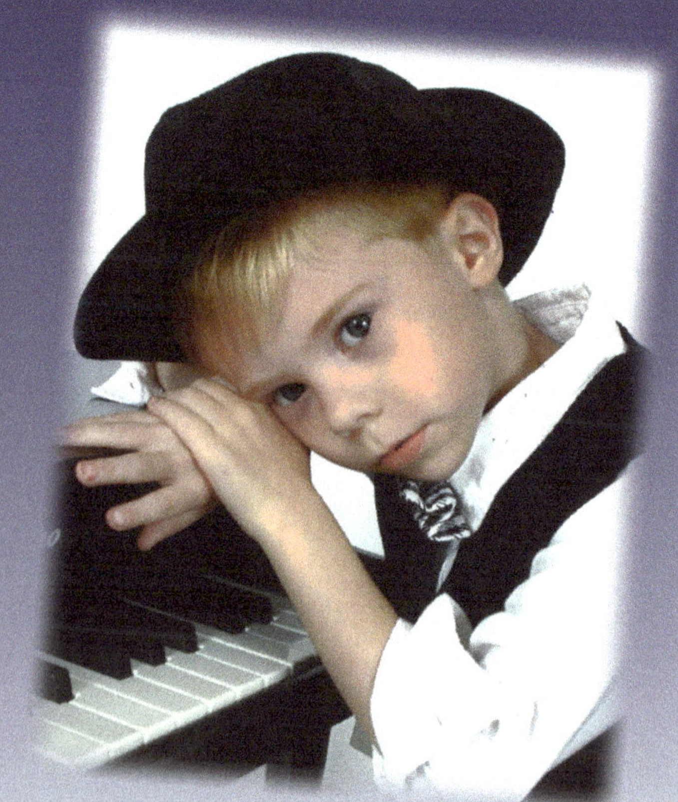

We are so proud of you Harrison!

LOVE ALWAYS and FOREVER,

**Mommy, Daddy,
Bauble, Papa,
Gran, Darin, Selena,
Austin, and Turner**

Brooke Wolcott

Duchess of Brookeland

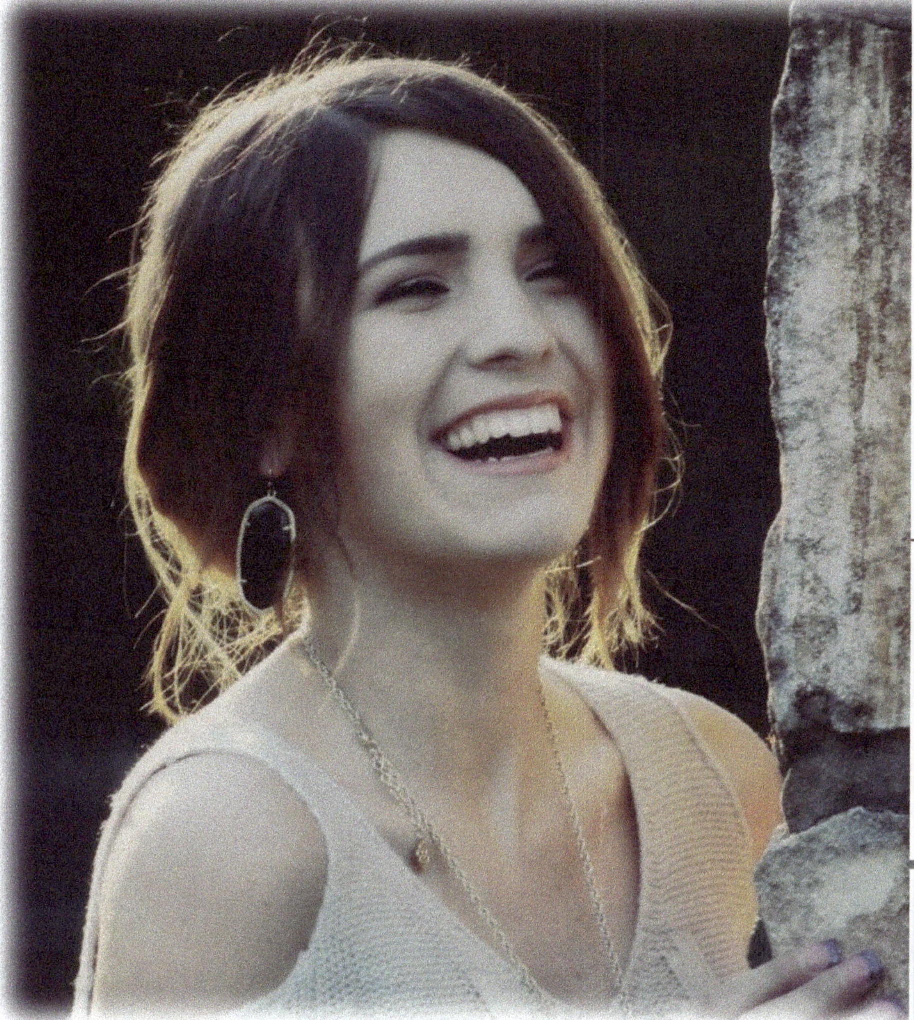

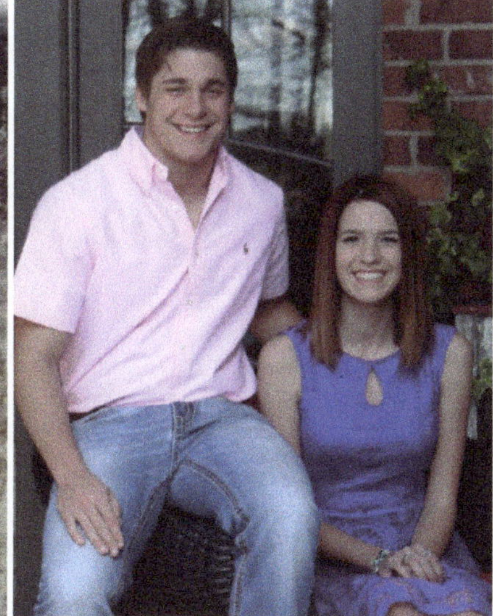

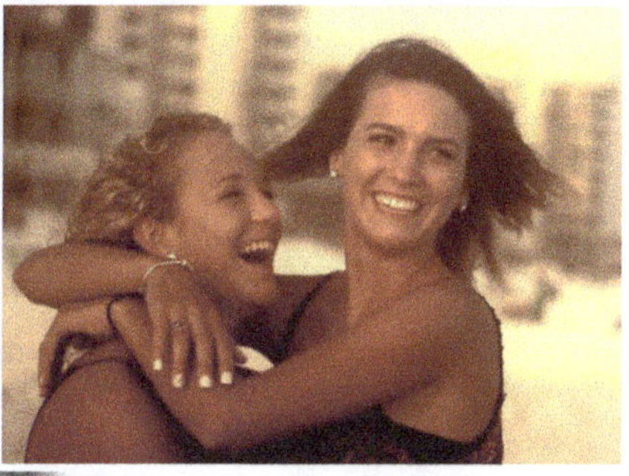

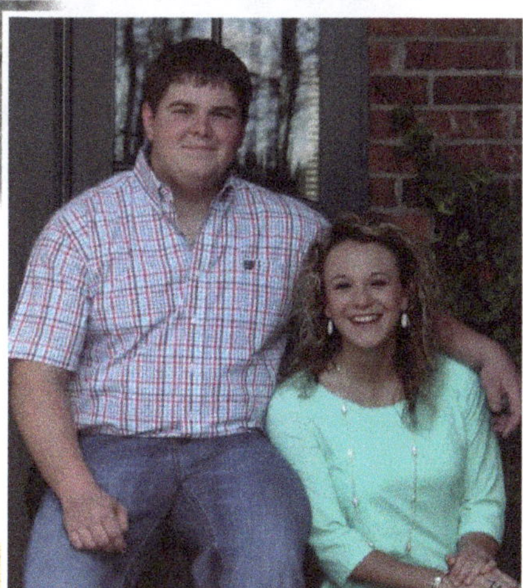

We're so proud of you!

Love, Mom and Dad

Compliments of

Wolcott & Associates ECS, LLC

1905 W. Bluff St. • Woodville, Texas 75979 • 409-331-9175

Erin Morgan Seamans
COLMESNEIL LADY-IN-WAITING

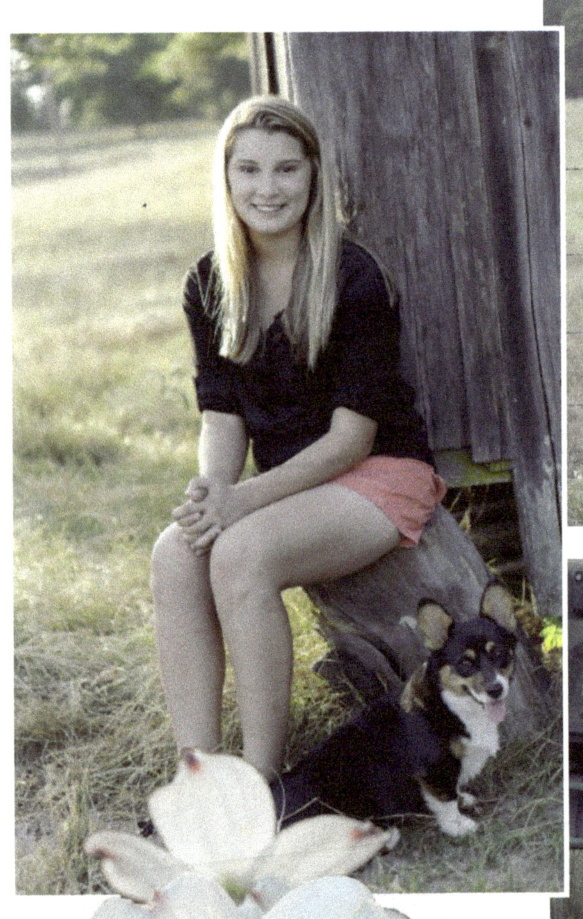

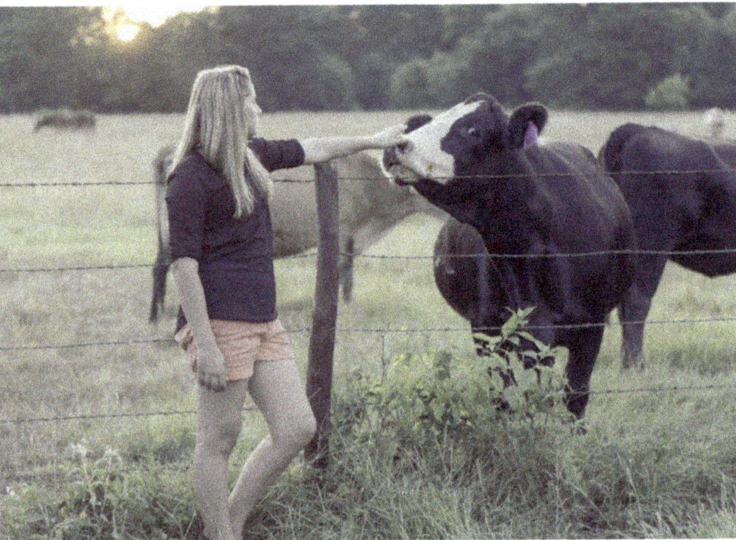

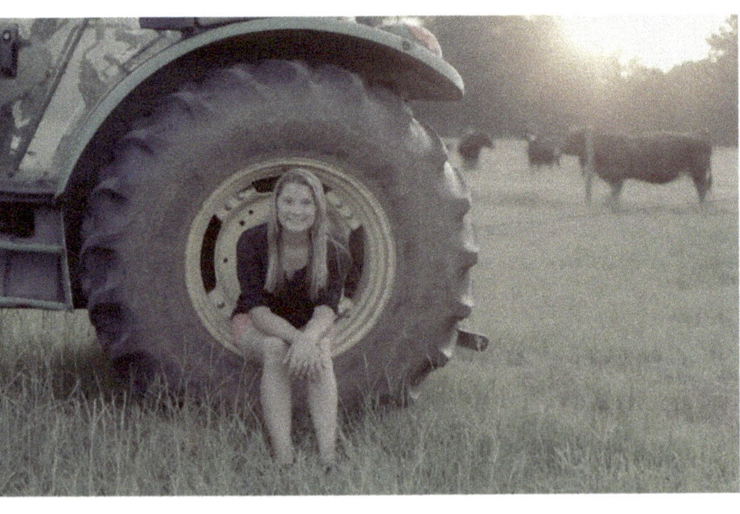

Enjoy the Dogwood Festival

These are the *Moments* to remember *Forever*

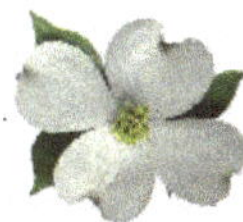

Previous Dogwood Queens

1940....................Lucy Jane Nellius	1966....................Amanda Haralson	1988....................Kathleen Mauer
1941....................Jacqueline Connally	1967....................Jeanette McClure	1989....................Tania Celestine
1946....................Helen Sheffield	1968....................Dell Jordan	1990....................Amy Craine
1947....................Patsy Walker	1969....................Kaye Owens	1991....................Angela Blakeney
1948....................Fredna Matthews	1970....................Jill Ogden	1992....................Whitney Lamberth
1949....................Ann Rainey	1971....................Robbie Crawford	1993....................Kimberly Woodrome
1950....................Nelva Hollomon	1972....................Jenny Whitworth	1994....................Jolea Glenn
1951....................Rosabell Sheffield	1973....................Charlotte Hancock	1995....................Lindsey Chapman
1952....................Bernice Phillips	1974....................Belinda Currie	1996....................Christine Heiman
1953....................Betty Riley	1975....................Nancy Nichols	1997....................Summer Norwood
1954....................JoAnne Young	1976....................Jill Crawford	1998....................Laura Kenesson
1955....................Marilyn Freeman	1977....................Cindy Hryhorchuk	1999....................Celeste Jordan
1956....................Seta Earl Triplett	1978....................Melinda Raiford	2000....................Brooke Brochtrup
1957....................Carolyn Platt	1979....................Susan Brodnax	2001....................Maegan Best
1958....................Karon Brodnax	1980....................Gina Temple	2002....................Sara Erdek
1959....................Beverly Standley	1981....................Allison Smith	2003....................Janna Kerr
1960....................Glenda Sanders	1982....................Cheryl Bass	2004....................Jessica West
1961....................Carol Ann Bennett	1983....................Ama Durham	2005....................Blaire Blanchette
1962....................Pamela Mills	1984....................Becca George	2006....................Payton Brown
1963....................Carolyn Clemmons	1985....................Joan Jordan	2007....................Melissa Garess
1964....................Linda Read	1986....................Monica Minter	
1965....................Terry Daniels	1987....................Pennie Glover	

2008
SHAYLA YOUNG

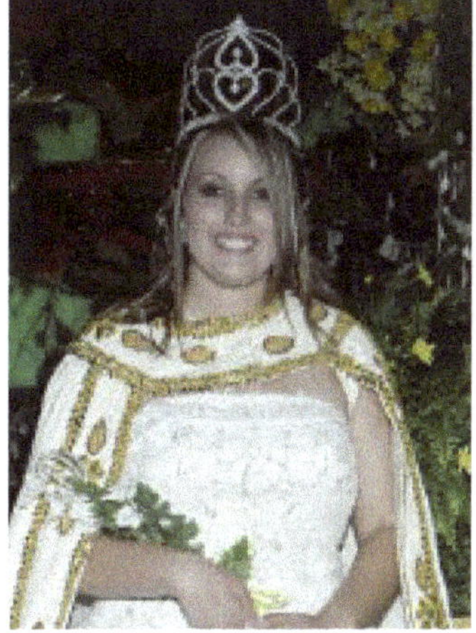

2009
ROSSI GARDNER

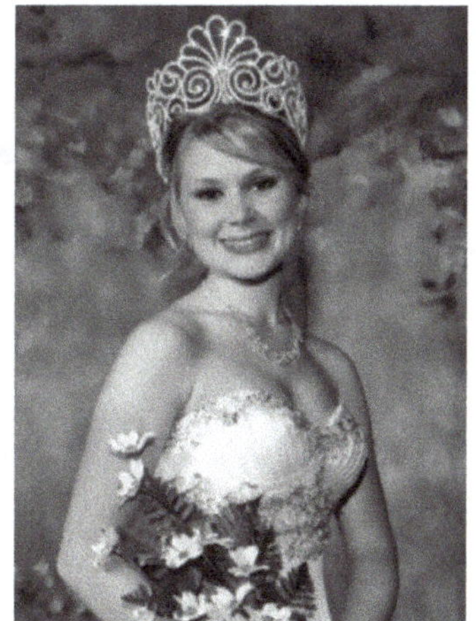

2010
KAYLEE RYAN

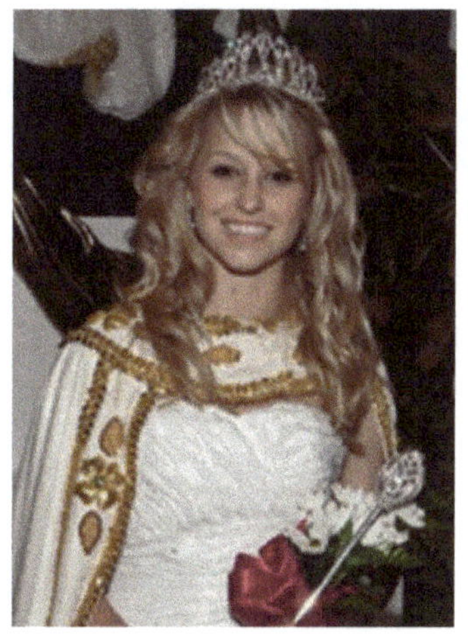

2011
KELSEY LANGHAM

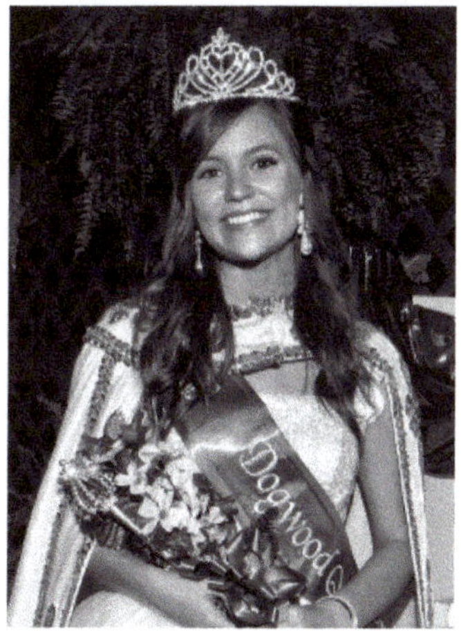

2012
ASHLEY LIVINGSTON

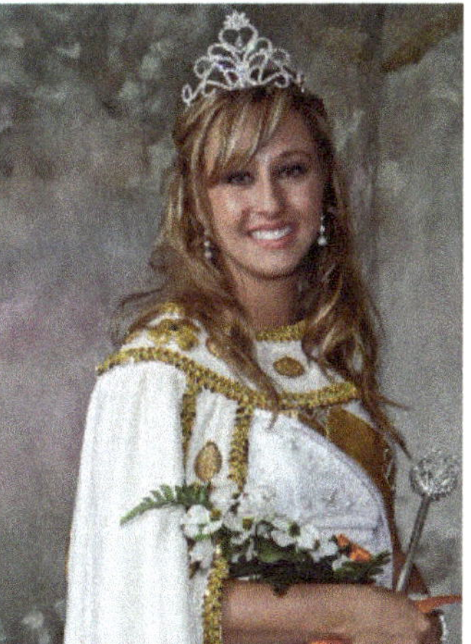

2013
MORGAN HAMMONS

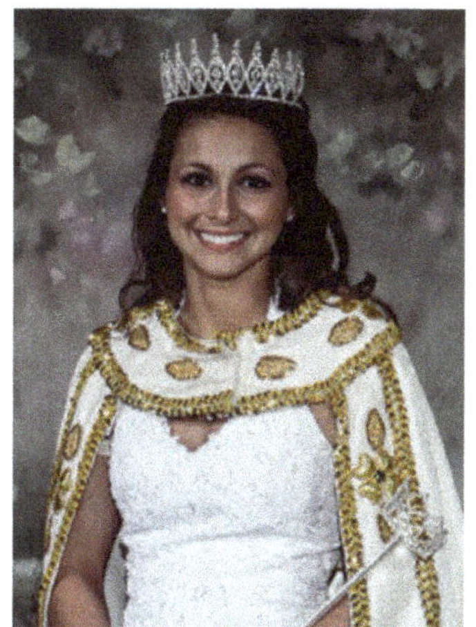

2014
DORIAN MALOY

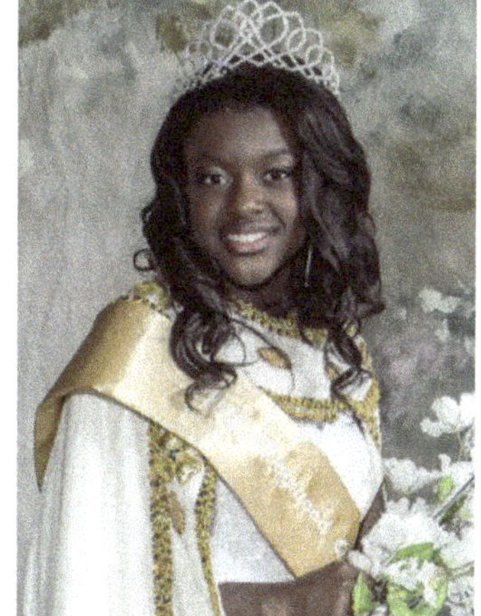

2015
LAKEN READ

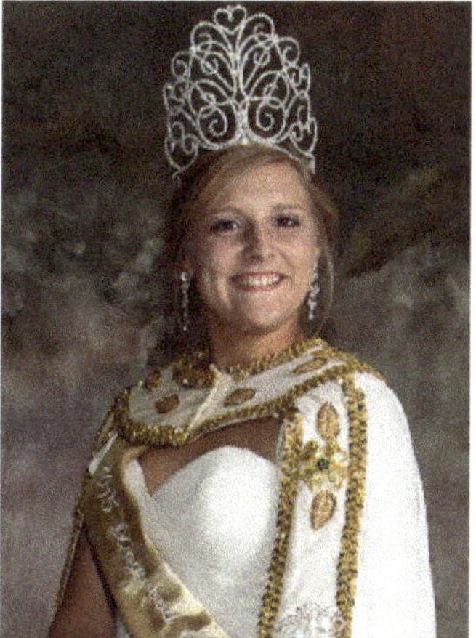

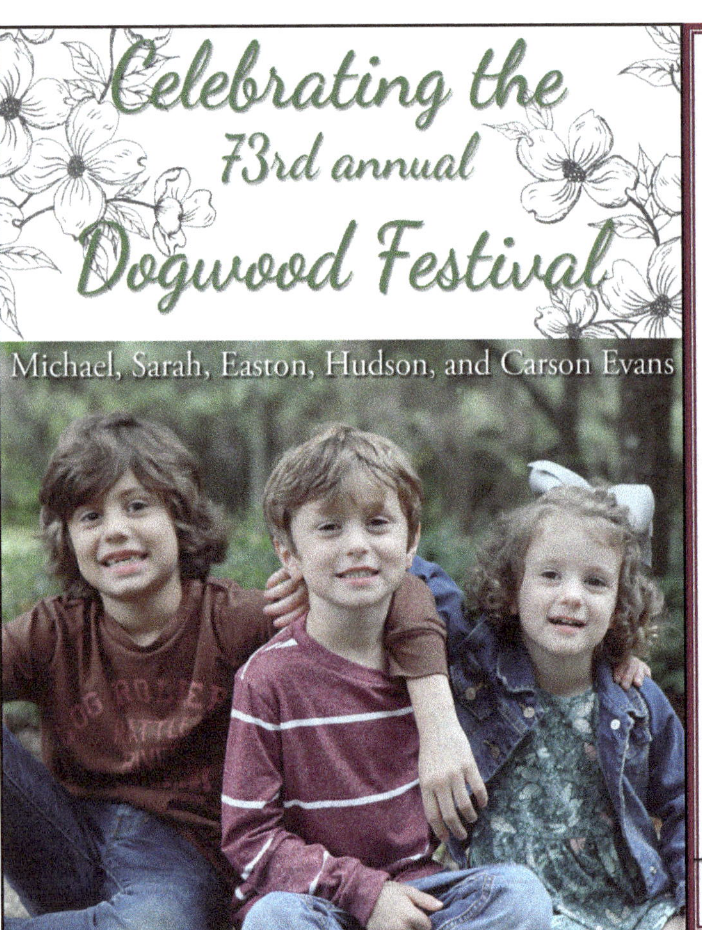

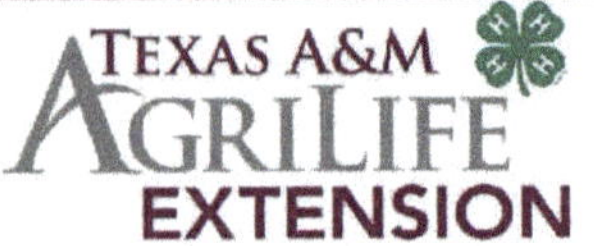

Tyler County AgriLife Extension Would Like to Congratulate and Wish the Very Best to the 2016 Dogwood Royal Court!

Kelly Jobe CEA FCS/4H
Jacob Spivey CEA Ag/NR
Maegan Odom Support Staff

Jacob.spivey@ag.tamu.edu ● Kelly.jobe@ag.tamu.edu

201 Veterans Way Woodville, TX 75979
Ph: 409-283-8284 Fax: 409-331-0015

Welcome to Tyler County!

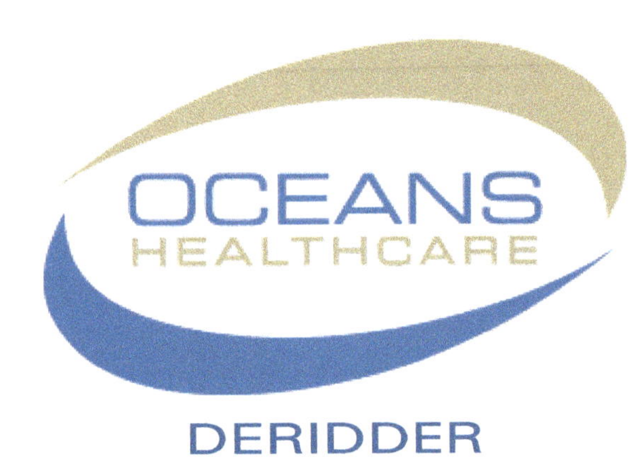

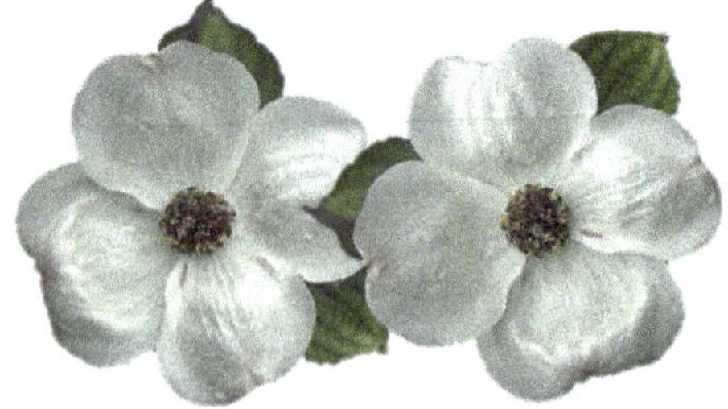

Proudly Serving the Residents of Tyler County for 10 Years

Tyler County Hospital

1100 WEST BLUFF   WOODVILLE, TX   409-283-8141

We Can Take Care of Your Emergencies, Hospitalizations, Labs and X-Ray Needs Right Here at Home

Tyler County Hospital is a licensed 49-bed acute care facility owned and operated by the Tyler County Hospital District.

The hospital provides a wide spectrum of healthcare services including a full-service emergency room and ancillary services such as laboratory, x-ray, ultrasound, CT scan, and Stroke Certification for all ER doctors and all TCH nurses.

The inpatient staff members provide skilled nursing care to a wide range of diagnostic entities, with the assistance of state-of-the-art cardiology equipment. Our staff pride themselves not only on the standards of care, but the personalized approach to patient care.

Patient Services

Laboratory
- *Hematology*
- *Coagulation*
- *Chemistry*
- *Serology*
- *Immunology*

Cardio/Pulmonary
- *EKG*
- *Pulmonary Function Test*
- *Stress Test*
- *Holter Monitors*

Imaging
- *Echo Cardiogram*
- *X-ray*
- *Fluoroscopy*
- *Ultrasound*
- *CAT Scan*

Other Services
- *Urology consultation, treatment & cystoscopy*
- *Cardiology Consultation, diagnosis & treatment*
- *Carotid & Cardiac Doppler Echo Studies*
- *Hearing Test*
- *Podiatric Medicine*
- *Rural Health Clinic*
- *LVN School*
- *Nephrology*
- *501c3 Foundation*

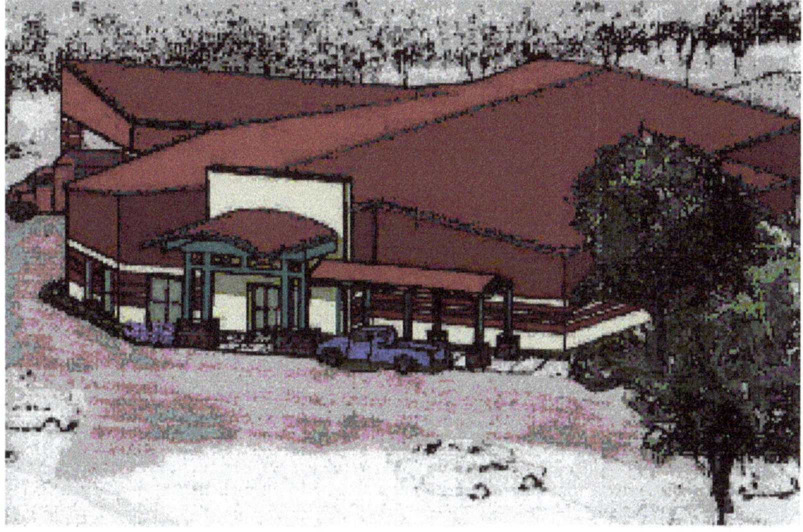

Tyler County Hospital has served the community for over 60 years. We have a Level 4 Trauma Center with state of the art equipment and we are regionally reviewed for stroke care. We have added PACS (Picture Archiving and Communication System) to our radiology department. Check with us to see which of your health care needs can be covered close to home.

Heritage Village Museum

157 PR 6000 • WOODVILLE, TEXAS

Owned & Operated by the Tyler County Heritage Society

PICKETT HOUSE RESTAURANT

TOUR THE VILLAGE

Self-Guided Tours • Gift Shop • Open Mon-Fri 9am-3pm, Sat & Sun 9am-5pm

PICKET HOUSE

Homestyle Cooking at it's Best • Mon-Fri 11am-2pm • Sat & Sun 11am-6pm

ANNUAL HARVEST FESTIVAL

Third Weekend in October

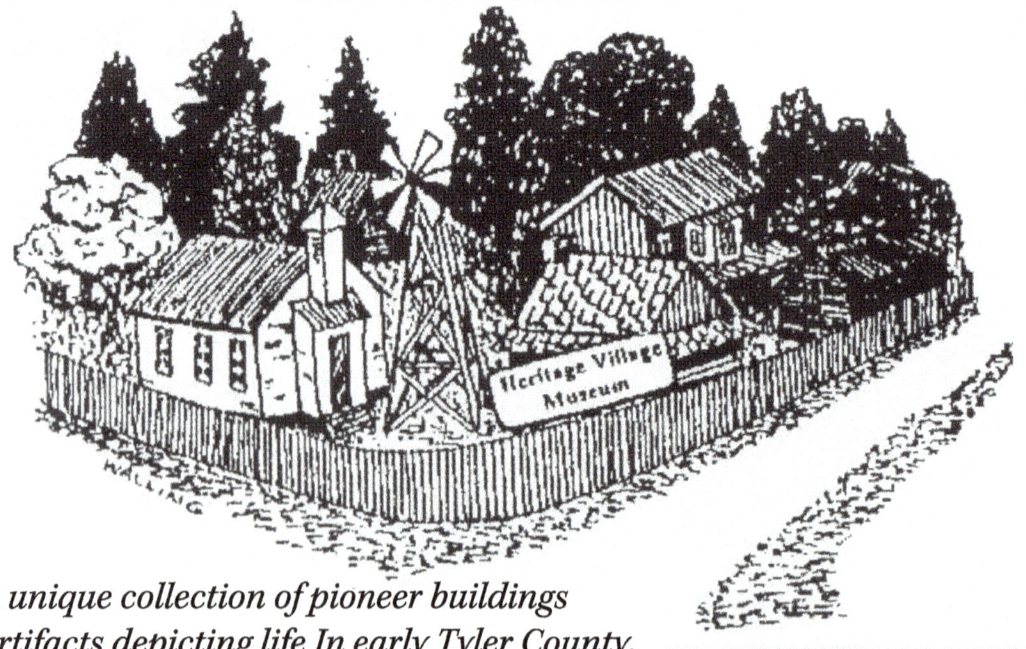

*A unique collection of pioneer buildings
and artifacts depicting life In early Tyler County.*

www.heritage-village.org • 409-283-2272

email: hvillagemuseum@att.net

Festival of the Arts Weekend

PHOTOS COURTESY OF
BOB BOYKIN PHOTOGRAPHY

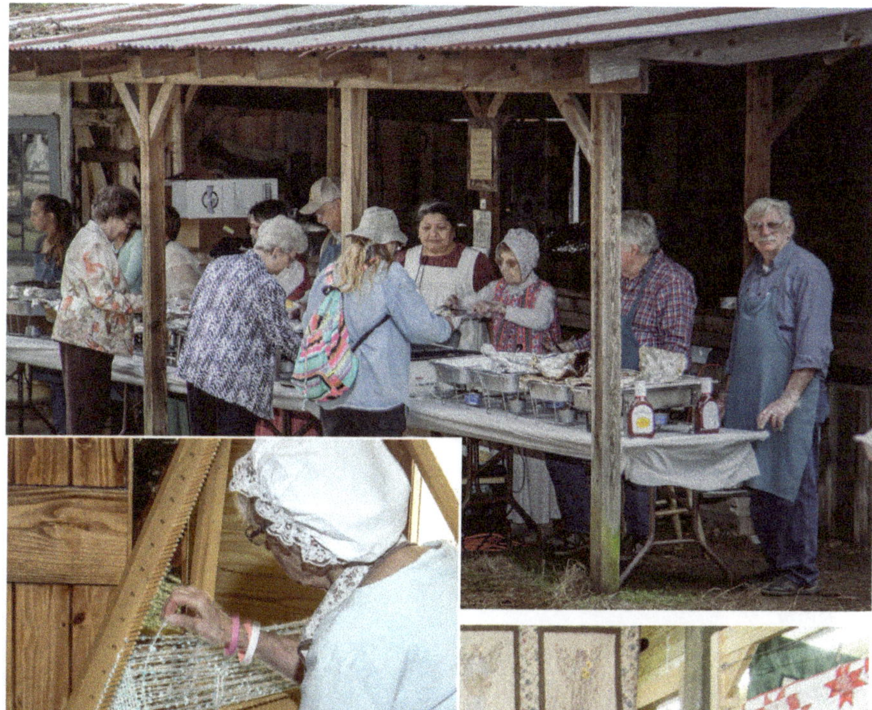

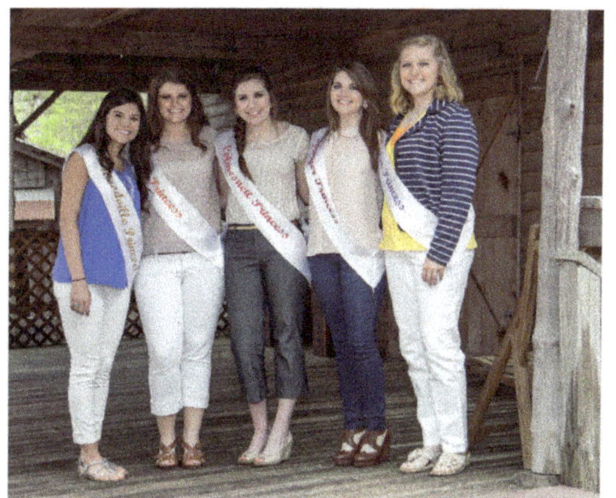

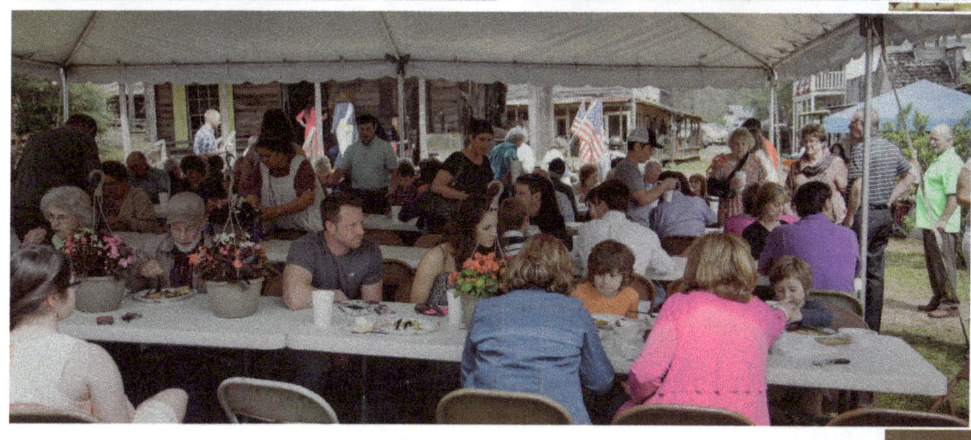

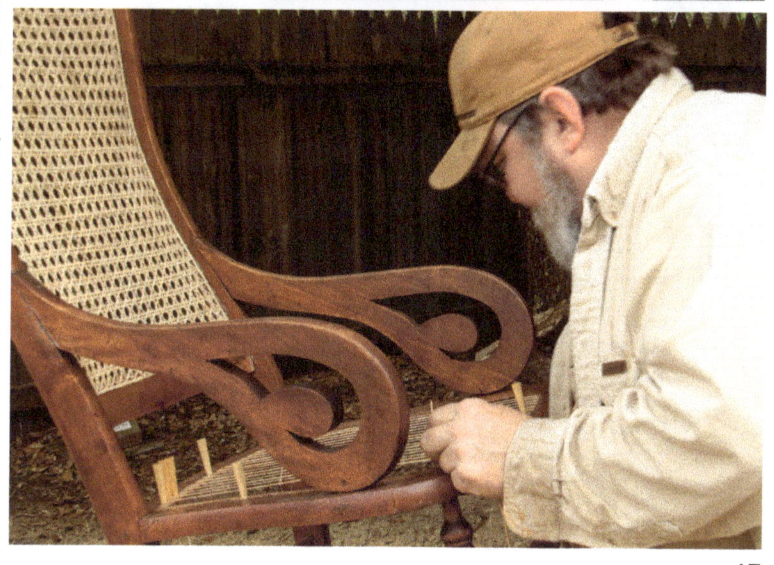

In March of 2000, the first Festival of the Arts Weekend was held at Heritage Village. The Dogwood Festival had always been a time for reflecting on our past, and it only made sense to look back at our heritage. Tyler County is very lucky to have Heritage Village, which offers a unique collection of historic buildings and artifacts depicting life in Tyler County from the 1860's to the 1900's. The Village features the 1866 Tolar Cabin, awarded the official Texas Historical Medallion, a reconstruction of the Z. C. Collier Store, once the hub of the thriving community of Town Bluff, the first county seat, and the Pickett House, which serves family-style meals in an old school house. Nature walks are also available.

Heritage Village is located one mile west of Woodville on Highway 190 and is a wonderful place to visit at any time of the year!

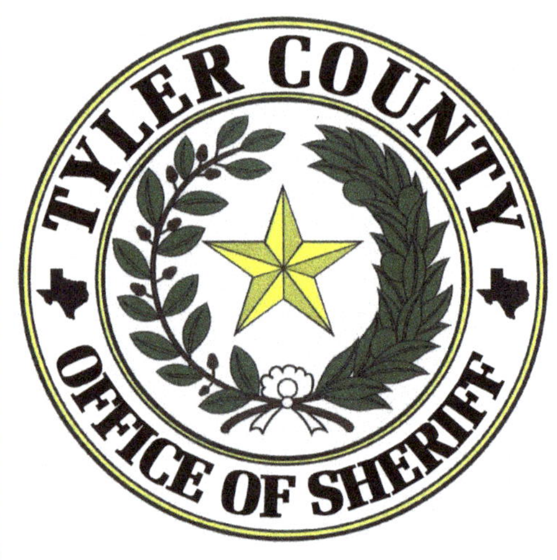

Wishing You a Safe & Enjoyable Dogwood Time In Tyler County

BRYAN WEATHERFORD
Tyler County Sheriff

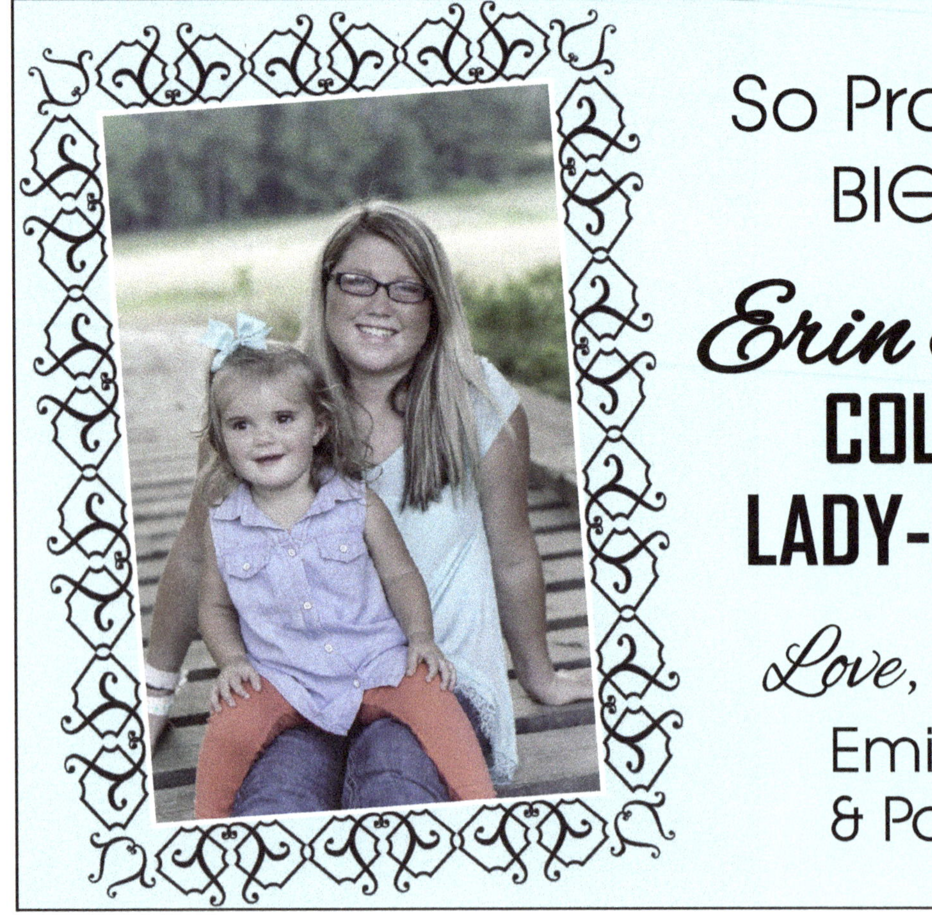

So Proud of Our
BIG SISTER!

Erin Seamans
**COLMESNEIL
LADY-IN-WAITING**

Love,

Emilee
& Paisley

Lindsey B. Whisenhant

ATTORNEY AT LAW

409-283-8288

Main Office Address:
130 South Charlton Street
Woodville, TX 75979

www.lbwlawoffice.com

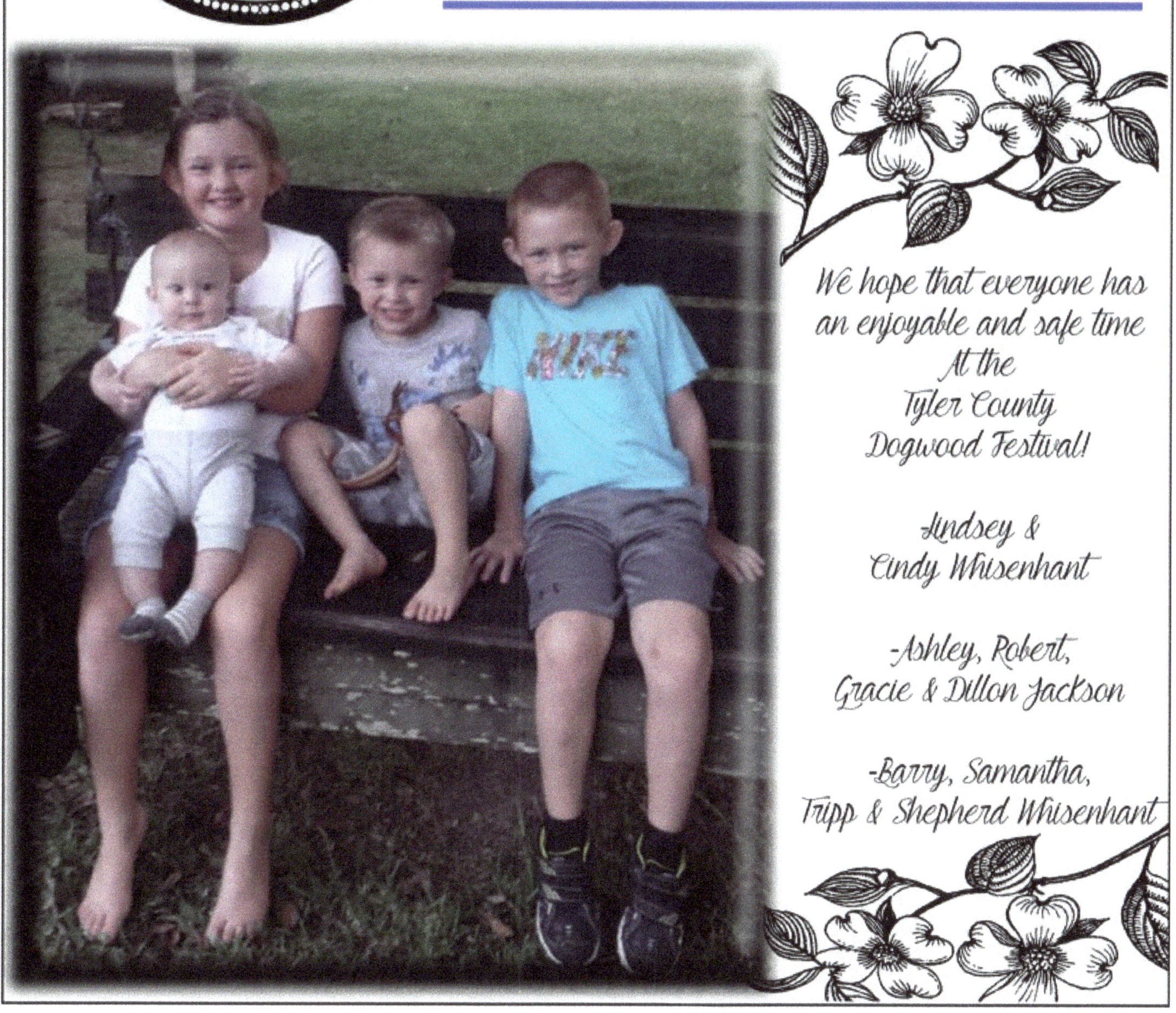

We hope that everyone has
an enjoyable and safe time
At the
Tyler County
Dogwood Festival!

Lindsey &
Cindy Whisenhant

-Ashley, Robert,
Gracie & Dillon Jackson

-Barry, Samantha,
Tripp & Shepherd Whisenhant

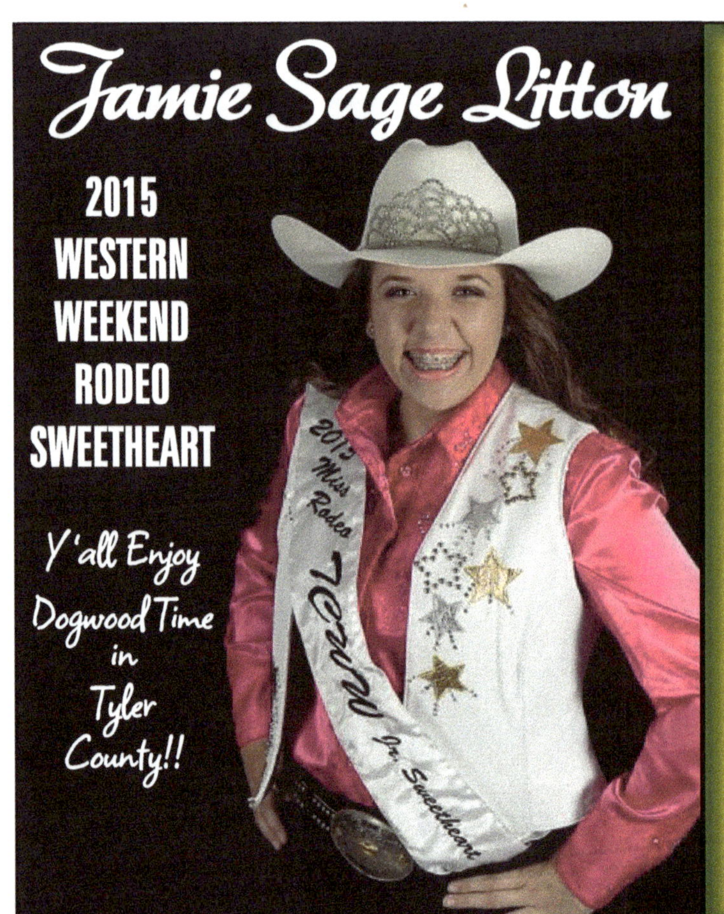

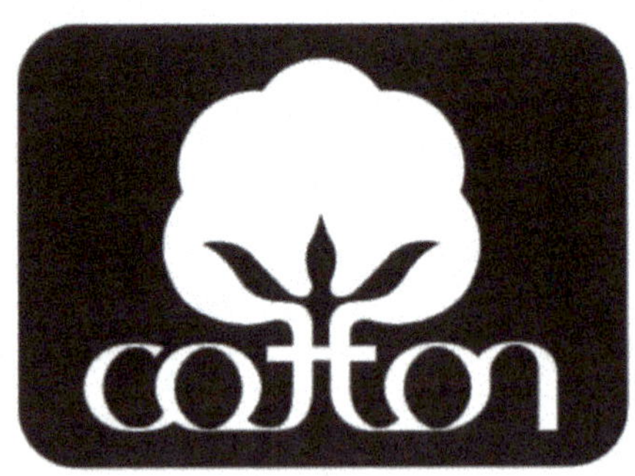

Compliments of

The McGehee's
Julie   Heath
Price   Ben   Drew
Rex   Meghan

IDALOU, TEXAS

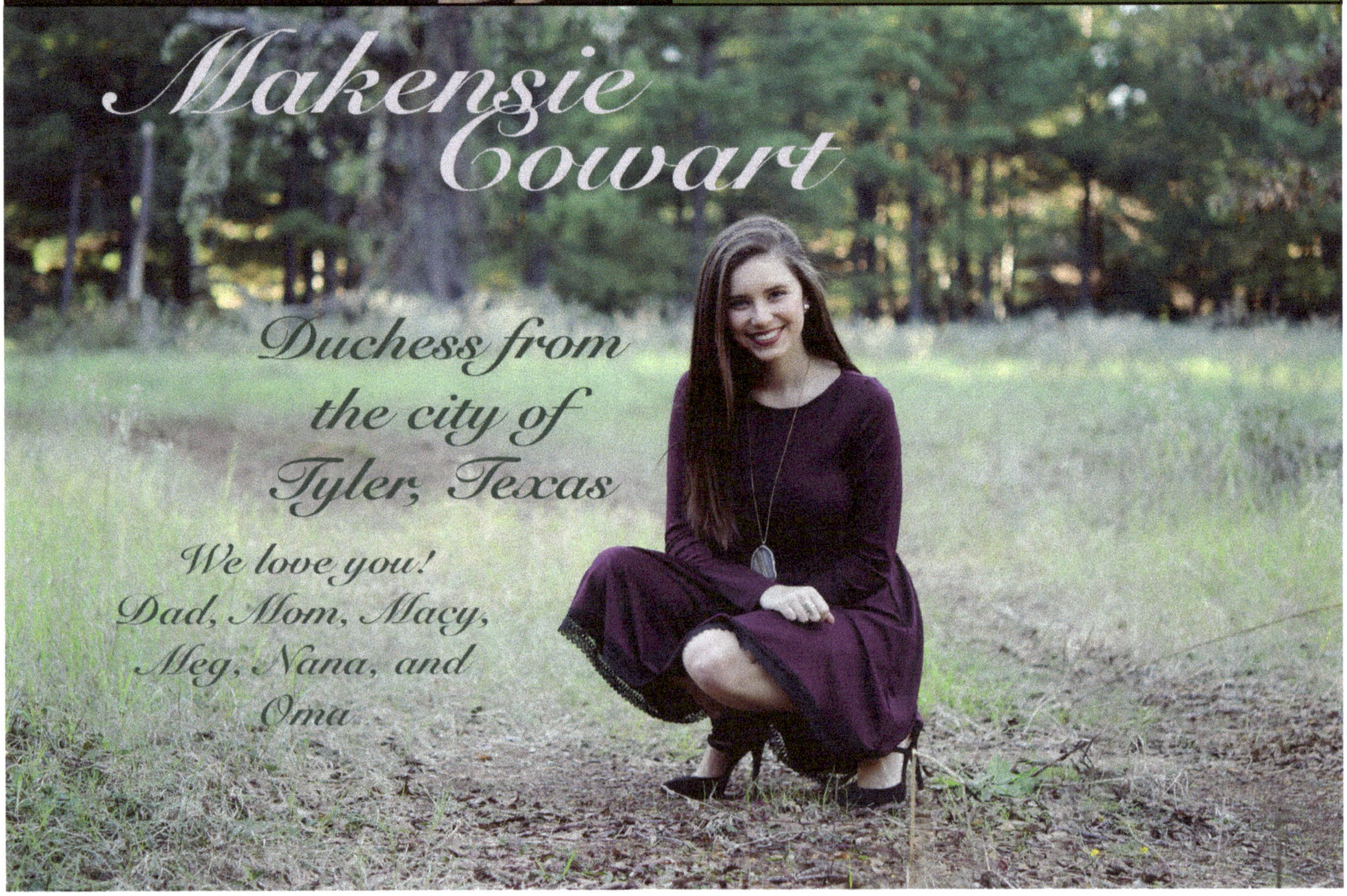

Western Weekend

In 1958 approximately seventy-five riders formed on Highway 190 West and rode into Woodville arriving in time to participate in the annual Dogwood parade. The western activities became so popular that during the 1960's the festival was expanded to include two weekends. In only eight years, Western Weekend had grown to include more than 1,800 horses and over 15,000 people coming to Woodville to enjoy a weekend of fun.

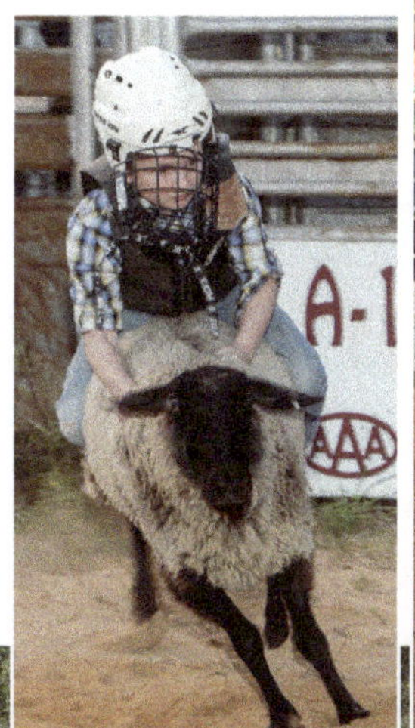

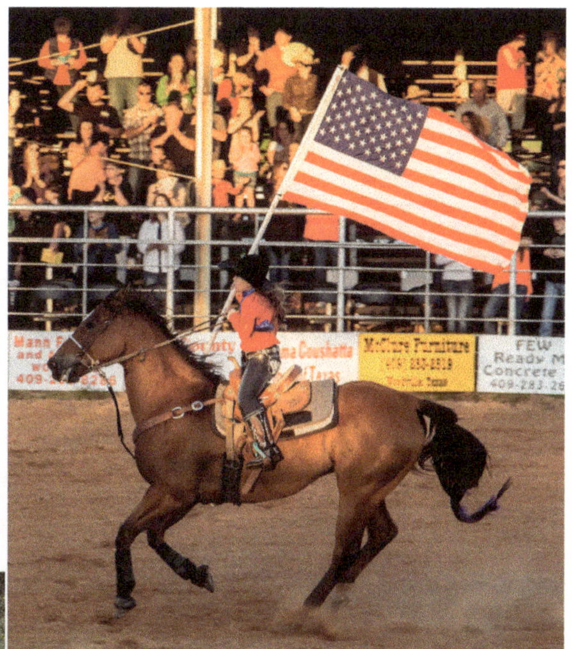

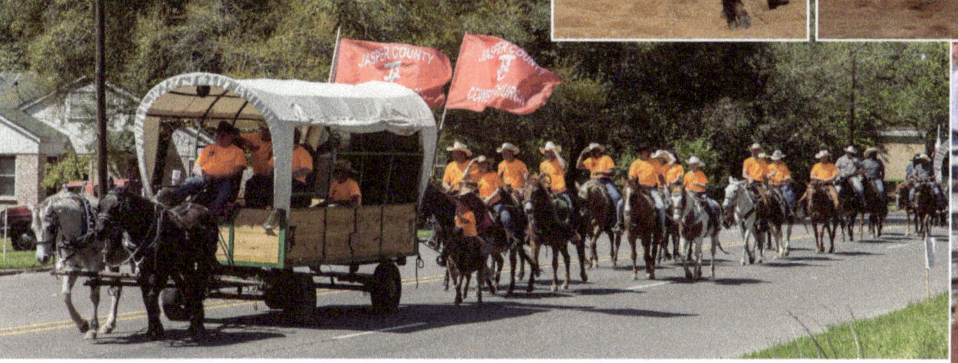

PHOTOS COURTESY OF BOB BOYKIN PHOTOGRAPHY

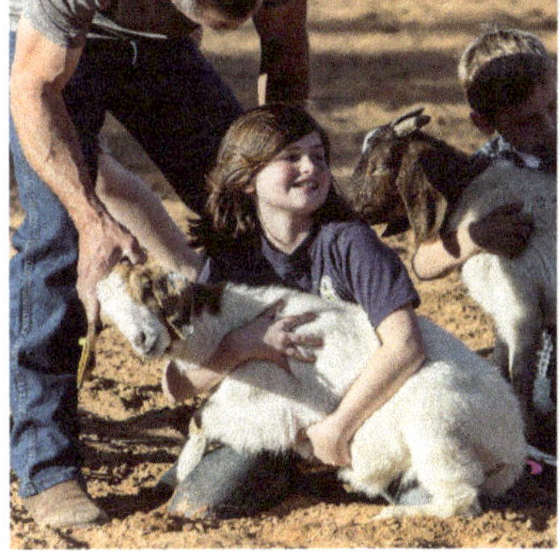

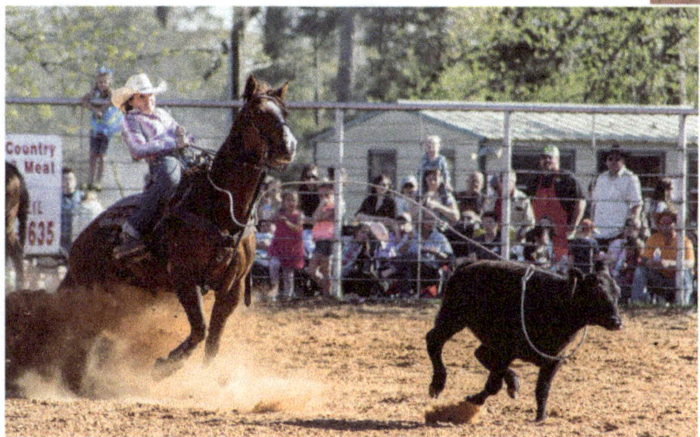

2015 TRAILRIDE WINNERS

Most Authentic: Southeast Texas Trailriders

Largest: B&J Ranch, Livingston

Best Dressed: Cowboys for Christ, Kountze

Close to 20 trailrides make their way to Woodville for the festivities. Each trailride stops along the way on Friday and makes camp. Western playdays and dances are held at this time so the weary riders can relax following the long day's ride.

OPEN RODEO

March 25 · 7:00pm
March 26 · 4:30pm

TYLER COUNTY FFA-4H ARENA

Bareback Riding
Mutton Bustin' · Calf Roping
Break-Away Roping
FFA Wild Donkey Saddling
Goat Scramble
Saddle Bronc · Team Roping
Barrel Racing · Bull Riding

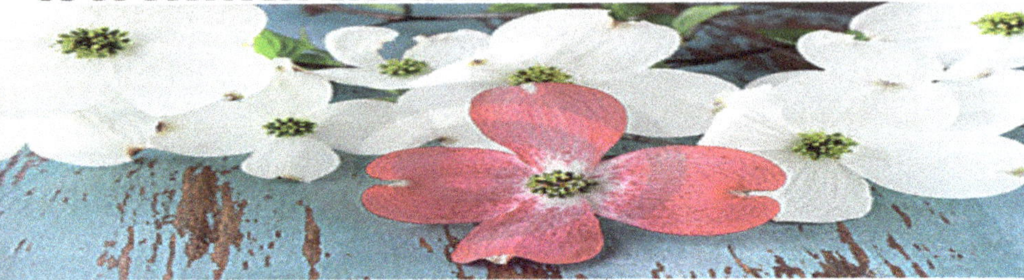

Compliments of

READ LOGGING INC.

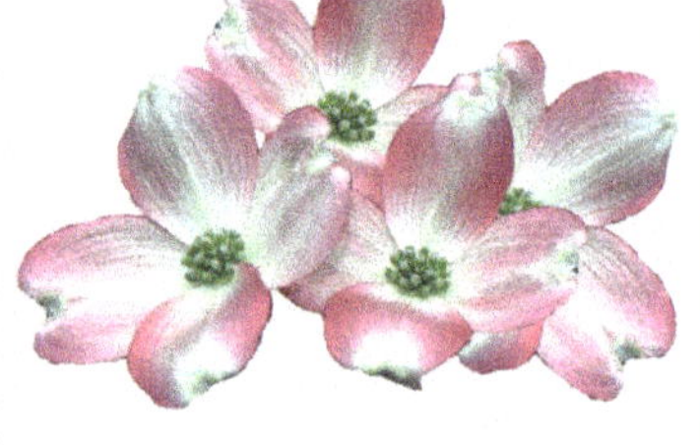

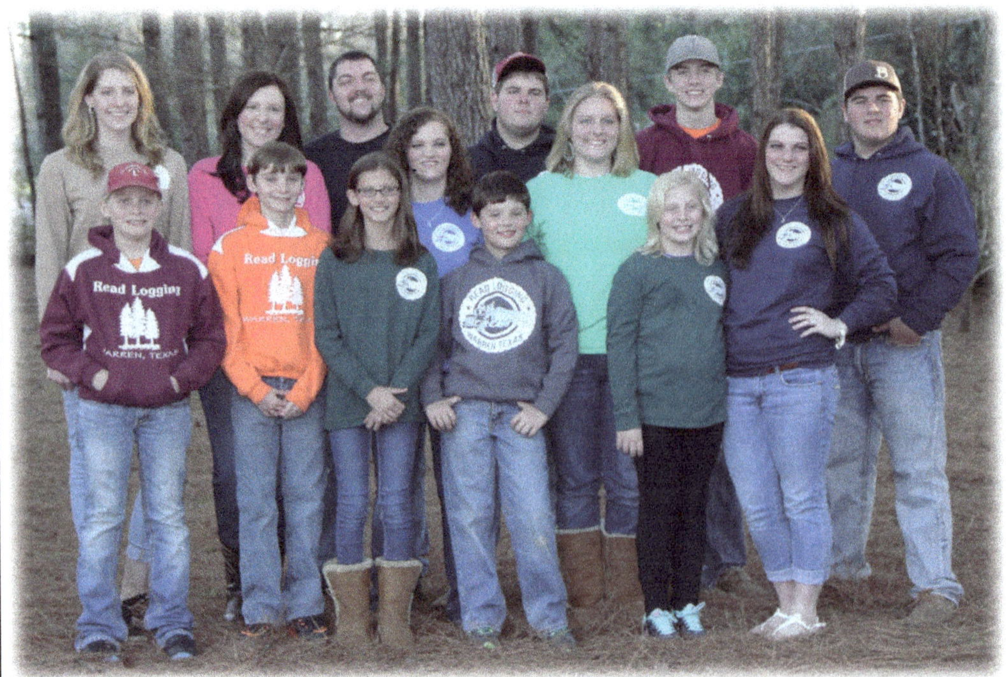

PATE'S COLLISION

Darion Pate, Owner • 409-283-6183
112 Cobb Mill Road, Woodville, TX 75979

*Proudly Supporting
The Tyler County
Dogwood Festival*

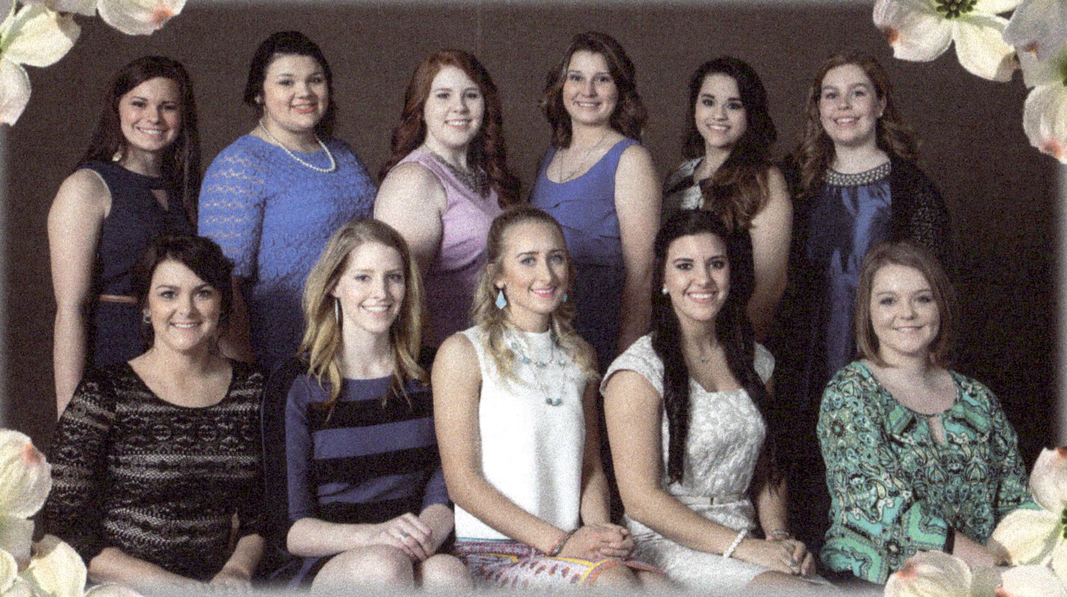

2016
Ladies-in-Waiting

2015 Western Weekend
Rodeo Sweethearts

Scarlett Addington
of Dayton, Texas

Western Weekend Sweetheart

Daughter of
Brett & Susan Addington

Jamie Litton
of Woodville, Texas

Junior Western Weekend Sweetheart

Daughter of Tracy Litton
and James Litton

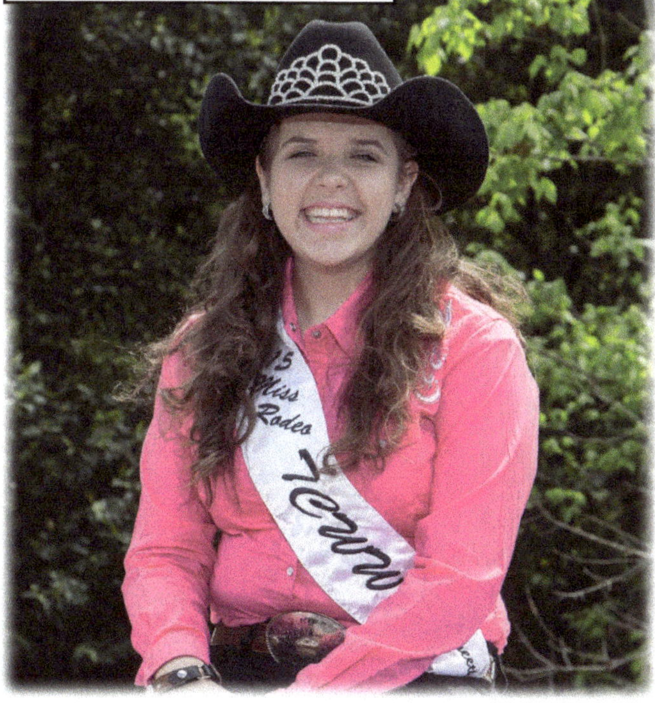

Previous Western Weekend Sweethearts

1972	Debbie Farmer
1973	Sherry Cassity
1974	Janet Martindale
1975	Cindy Jordon
1976	Keitha Massey
1977	Melissa Boles
1978	Toni Stephenson
1979	Kerri Cassity
1980	Jana Lin Alexander
1981	Donna Owens
1982	Melissa Kahla
1983	Tanya McInnis
1984	Tonni Hawthorne
1985	Monte Conn
1986	Shaunde' Parker
1987	Amber Bell
1988	Charlotte Kahla
1989	Misty Hoyt
1990	Brandi Gloor
1991	Kendra Rosenberger
1992	Dwawn Childers
	Emily Spurlock
1993	Lindsey Chapman
1994	Donna A. Stricklen
1995	Dawn Licatino
1996	Victoria Gordon
1997	Amber Carpenter
1998	Jill Larrison
1999	Sarah Havens
2000	Jennifer Carter
2001	Karissa Chapman
2002	Rachel Oates
2003	Allison Bray
2004	Joanna Blackwell
2005	Alyssa Plackemeier
2006	Kira Knaupp
2007	Rachell Rathmell
2008	Danielle Kemp
2009	Garen Reese
2010	Victoria Welch
2011	Rosanna Price
2012	Carrie Wilbert
2013	Natalie Moffett
2014	Courtney Elliott
2015	Scarlett Addington

Previous Junior Sweethearts

1997	Stormi Marie Guidry
1998	Suzanna Sutoon
1999	Amy Woods
2000	Chelsea Beggs
2001	Josie Robinson
2002	Ashley Chambliss
2003	Salina Parker
2004	Shelby Phillips
2005	Lauriann Beggs
2006	Miranda Allen
2007	Hallie Smith
2008	Natalie Moffett
2009	none
2010	Sissy Winn
2011	Haley Hensarling
2012	Emily Andrus
2013	Abbie Suggs
2014	Cheyanne Barrett
2015	Jamie Litton

SWEETHEART DONORS

Buckles:
East Texas Home Health

Other Contributors:
Napco,
Citizens State Bank,
Sullivan's Hardware,
JD Taylor Contractor

COLMESNEIL COMMUNITY CENTER

Reserve for your next event!

409-837-5211

Birthdays · Anniversaries · Weddings · Reunions
Meetings · Showers · Other Special Occasions

FULL KITCHEN · FIREPLACE
BASKETBALL, VOLLEYBALL & PLAYGROUND FOR KIDS

Proud Sponsor of the Tyler County Dogwood Festival!

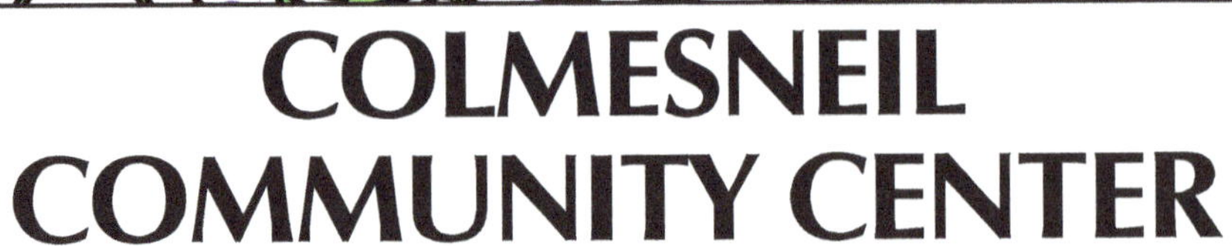

1986
Monica Minter
Gunter
Dogwood Queen

2014
Sarah
Gunter
Royal Duchess

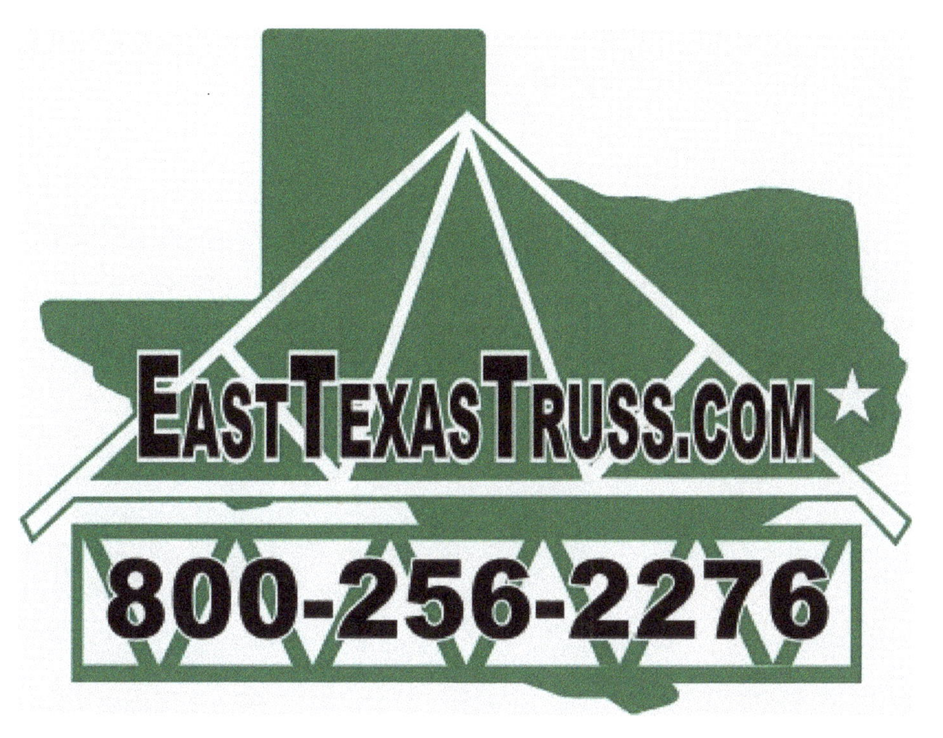

EastTexasTruss.com
800-256-2276
Building
Floors
and
Roofs
of the
Future
Today!
Supporting
TYLER COUNTY
and the
DOGWOOD FESTIVAL
Since 1988 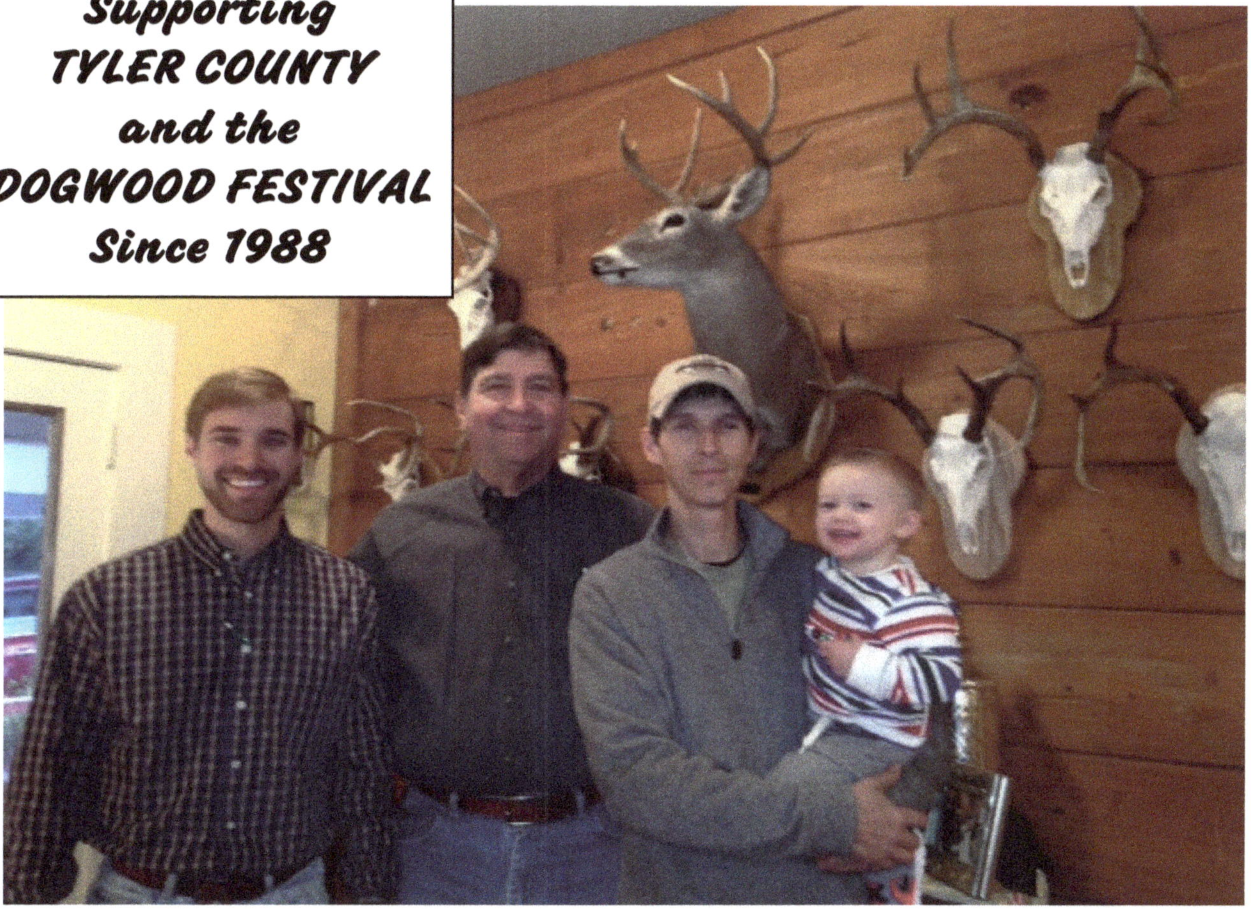
532 FM 1013, Hillister, TX 77624 • 409-283-3728

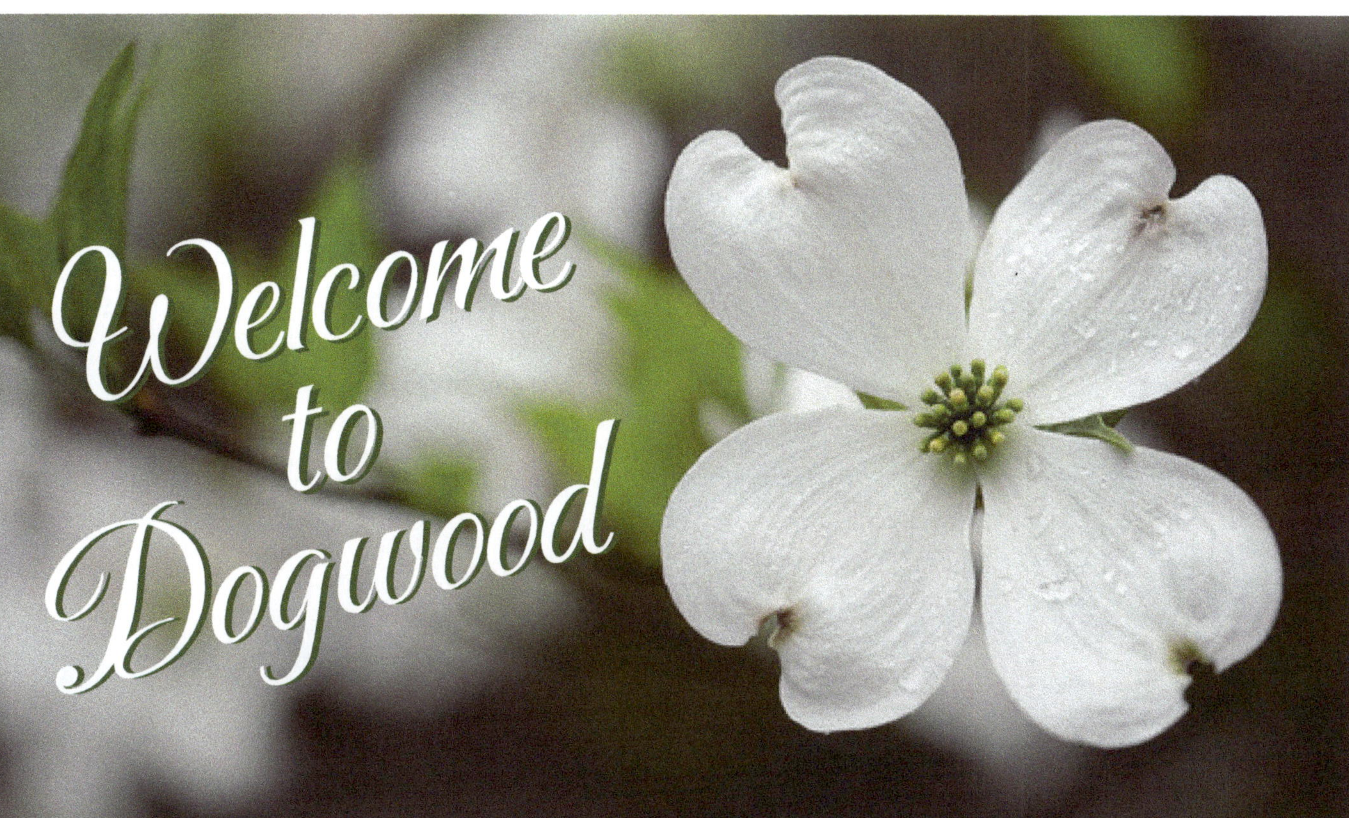

Compliments of

Tubb Investments

211 South Magnolia
Woodville, TX 75979

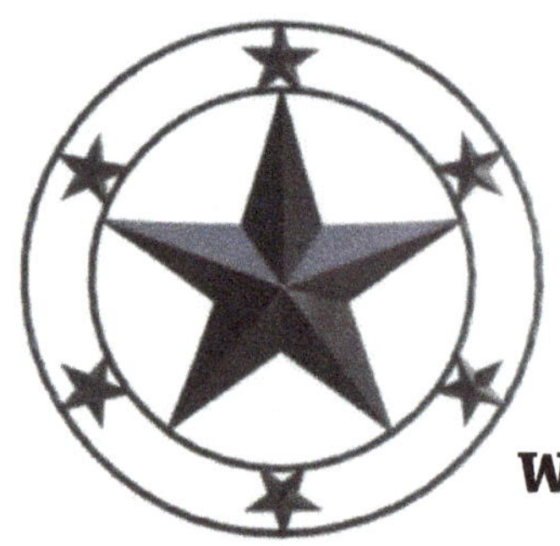

McClure Furniture Company

Complete Home Furnishings

**605 S. Magnolia
Woodville, Texas 75979**

409-283-2519

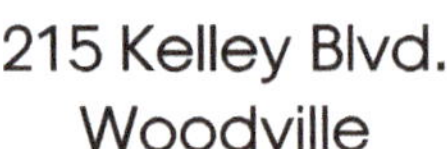

Woodville Florist & GIFT SHOP

215 Kelley Blvd.
Woodville

409-283-2571 • 1-800-548-4661
www.woodvilleflorist.com

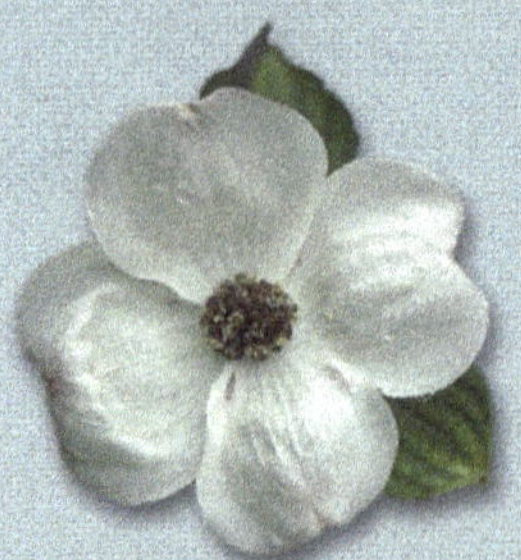

Kassidy Haralson

2016 Dogwood Festival Trainbearer

We love you Kass!

Mom,
Bryan,
Liam,
& Finn

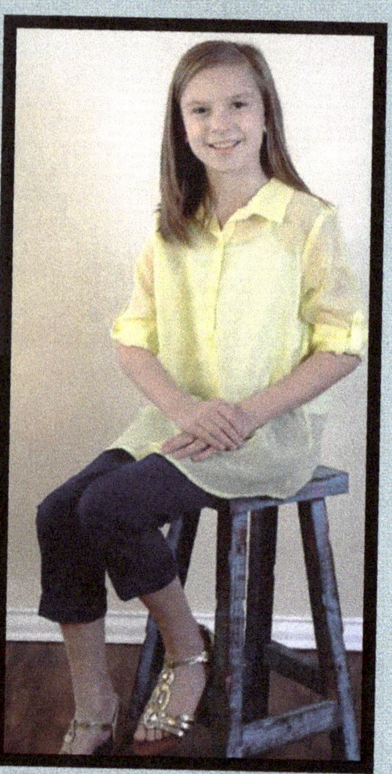

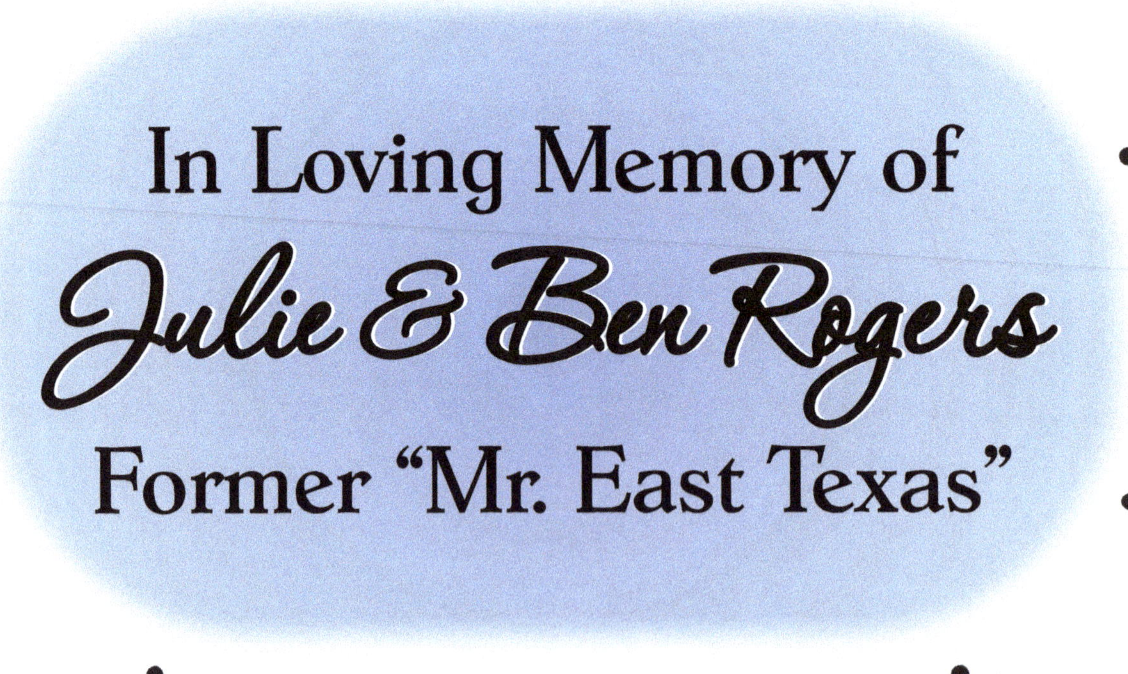

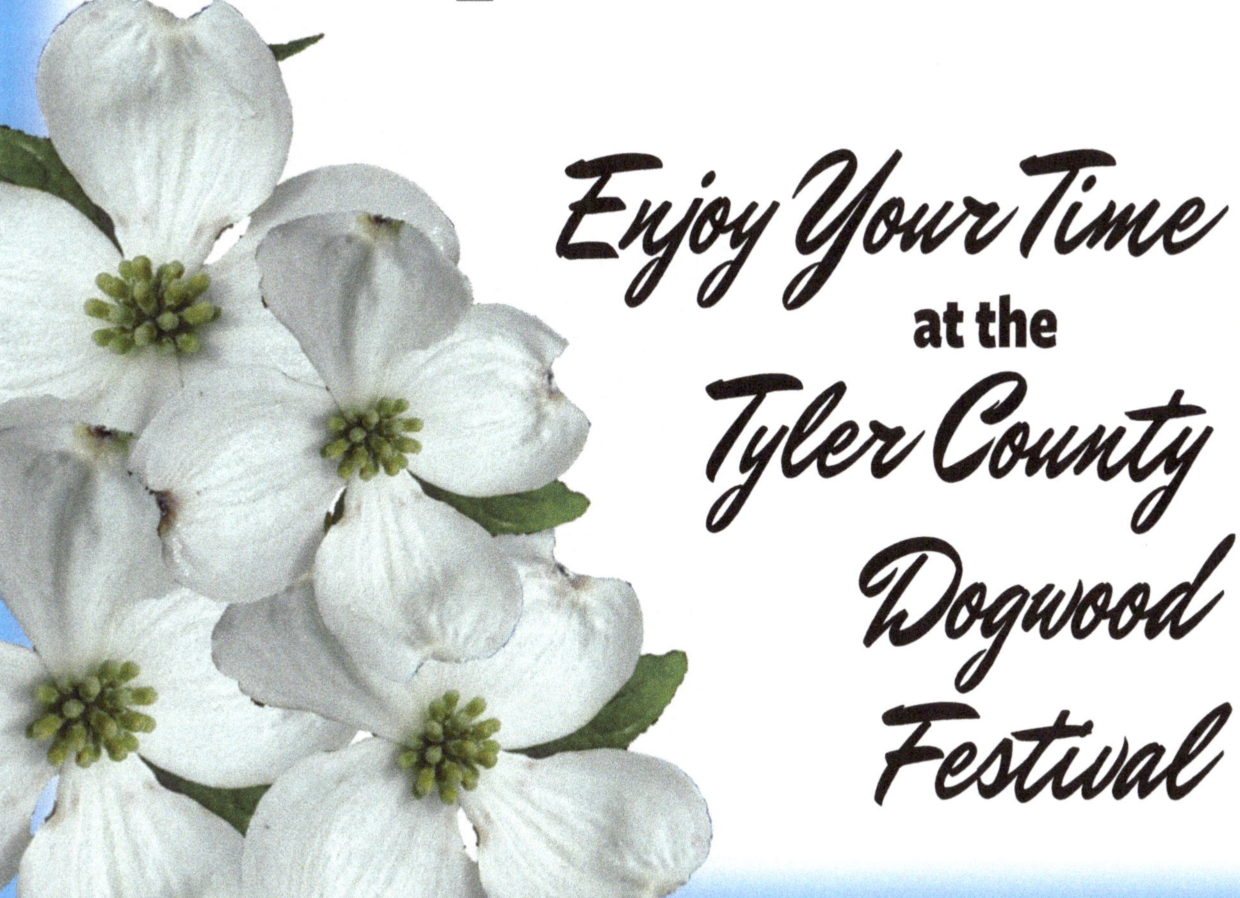

Compliments of
Tyler County
Industrial
Corporation

Enjoy Your Time
at the
Tyler County
Dogwood
Festival

DOGWOOD DIRECTORS

FESTIVAL PRESIDENT

Kim Shaw

PERMANENT DIRECTORS

FRED SULLIVAN	GLEN CONNER
GEORGE JARROTT	BELINDA ALLISON
JUDY BROWN	SUSAN BRODNAX AVRIETT
KATHERINE BRANCH	KIM OVERSTREET-SHAW
ERLINE INGLE	

FORMER EXECUTIVE DIRECTORS

JOSIAH WHEAT 1959-1971

FRED SULLIVAN 1972-1976

GEORGE JARROTT 1977-1979

JIM CLARK 1980-1985

BEN WORTHY 1986

BETSY STAFFORD 1987-1988

BECKY JINNETTE 1989-1993

JUDY BROWN 1994-1997

ELAINE TUBB 1998

KATHERINE BRANCH 1999-2003

BECKY JINNETTE 2004-2005

KIM OVERSTREET-SHAW 2006-2012

BUCK HUDSON 2013-

KINGSMAN

Bob Boykin, Dr. James Brown, Kirk Brown, Chase Burkhalter, Glen Conner, John Cooley, Joseph Conner, Michael Evans, Steve Evans, Keith Hays, Stephen Hill, Dwayne Hollingsworth, Buck Hudson, Hale Hughes, Cody Jarrott, George Jarrott, Fulton Jeansonne, Mark Johnson, Robert Lebouef, Walter McAlpin, Eric Narvaez, Curtis Pittman, Billie Read, John David Risinger, Phillip Scoggins, Lance Seamans, Don Shaw, Jacob Spivey, Kyle Spivey, Dr. Caleb Spurlock, Fred Sullivan, Billy Tolar, Casey Tolar, Bryan Weatherford, John Wilson, Morgan Wright, Jim Zachary

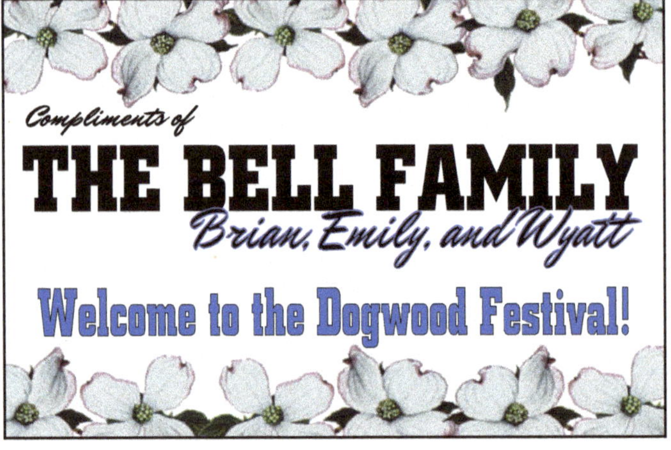

OPERATING DIRECTORS Seated left to right: Chesney Wright, Christy Cooley, Leann Monk, Ofeira Gazzaway. Standing left to right: Alicia Scoggins, Kathy Carruth, Buck Hudson, Danita Skinner, Mandy Risinger Not pictured: John Cooley, Fulton Jeansonne. Carrie Standley

EXECUTIVE DIRECTOR

Buck Hudson

PUBLICATIONS DIRECTOR

Danita Skinner

Committee Members:
Bob Boykin, Hale Hughes, Mandy Livingston, Karen Hughes

FINANCE DIRECTOR

Christy Cooley

Committee Members:
Rebecca Hill

Dogwood Dash:
Sponsored by Citizens State Bank, Co-Chaired by Bernice Coats & Tami Powell

ENTERTAINMENT DIRECTOR

Alicia Scoggins

Royal Tea, Western Dance, Scavenger Hunt, and Ice Cream Social Committee Members:
Tanya Crawford Gill, Aubrey Citrano, Courtney Marshall-McHenry, Tiffany Borel, Harlie Fortenberry, Dottie Watts, Victoria Scoggins

Former Queen's Reception:
Robbie Evans, Jill Davis

Jean Fling:
Texas Business Women

Kingsman's Ball:
Tyler County Chamber of Commerce

Parent's Reception:
Woman's Study Club

Queen's Court Brunch:
Woman's Reading Club of Woodville

Royal Tea: Sponsored by First National Bank of Jasper

and COMMITTEE MEMBERS

DECORATIONS DIRECTOR
Chesney Wright

Throne Committee:
Ashley Burkhalter, Tara Burman, Sally Carlton, Emily Carney, Tori Harris, Amy Risinger, Amee Scarberry, Jamie Weaver, Ashley Weatherford, Heather Wood, Morgan Wright

Paper Flowers Chairperson: Danielle Speaks

Flower Petal Cutter: Danielle Speaks

Flower Making and Tying On:
Danielle Speaks, Beth Allison, Josh Allison, Sarah Allison, Lesha Burkhalter, Pat Collier, Helen Grammer, Amy Risinger, Patti Tucker, Rebecca Wiseman, Buck Hudson & Spurger Ag Floral Design Class, Jennifer Pierson & Woodville Intermediate Art Classses, Reagan Ivey & Woodville HS Ag/Floral Design Class, Katrina's Salon - Katrina Ferguson, Leah Harris, Cheryl Marshall, Cheryl Wardlaw, the Ladies of The Orchard, Woman's Reading Club of Woodville.

Greenery Gatherers:
Danielle Speaks, Buck Hudson & Spurger Ag Classes

Special Thanks:
To Sheriff Bryan Weatherford and the Tyler County Sheriff's Department for their assistance in putting out the flowers and cleaning them up each year!

WESTERN WEEKEND DIRECTOR
Kathy Carruth

Queen's Contest:
Sammie Seamans, Kathy Carruth, Heather Seamans

Western Parade:
P.J. Wade, Phillip Scoggins, Alan Gartnu

PUBLICITY DIRECTOR
Mandy Risinger

Ticket Sales Chairman: Sharice Allison

Ticket Sales Committee: Trey Allison, John & Cindi Cooley

Website: Daisy Marino

Former Queen's Information: Payton Clotiaux

Highway Banner: Dianne Lewis, Bubba Moore

Committee Members:
Laren Allen, Marsha Watts, Stephanie Voth, Joanna Jarrott, Holly Shirley

Videography: Woodville High School

PARADE DIRECTOR
Fulton Jeansonne

Float Committee:
Sharon Jeansonne, Emma Jeansonne, Julia Jeansonne, Pat Collier, Helen Grammer, Christina Drake, Trisha Brackin

Parade Committee:
Bill Brackin, Jason Drake, Terry Riley, Billy & Krystal Tolar

ROYALTY DIRECTOR
Carrie Standley

Committee Members:
Kimen Fondren, Tina Pittman, Michelle Tucker, Patti Tucker, Sarah Evans, Janna Kerr, Jana Rayburn, Joanna Jarrott

Queen's Court: Stephanie Read

Duchess Chairman: Candace Spivey

Children's Court Chairman: Sarah Evans

Children's Court Committee:
Buck Hudson, Candace Spivey, Danasa Rawls, Kimberly Parker, Katherine Branch, Patti Tucker, Joan Conner

Trainbearer's Chairman: Michelle Tucker

Trainbearer's Committee:
Patti Tucker, Candace Spivey, Stephanie Read, Ju Ju Standley, Danasa Rawls, Buck Hudson, Joan Conner

Coronation Flowers: Jenny Lewis

Donation of Sashes:
Given in honor of Annette Smith and Allison Smith Matthews by Rodney, Stephanie, & Carly Smith.

FESTIVAL *of the* ARTS DIRECTOR
Ofeira Gazzaway

Committee Members:
Charles & Wanda Smith, John Gazzaway, Carol Shields, Gloria Newsome, Blake Goss - Railroad Museum, Sassy Scrapper's Quilt Guild, Spinners & Weavers

HISTORICAL PLAY DIRECTOR
Leann Monk

Choreography: Rachel Hadnot

Stage Direction:
Pam Lewis, Joanna Jarrott, Rebecca Hill, Kristy Martin, Jana Rayburn

Ushers: Woodville Lions Club

Sound: Sound Techs, Lufkin, TX

Lights: Rodney Monk, Teddy Heckman

Costumes & Props:
Diane Lewis, Chester FFA

Set Design & Construction: Coty Squires, Rodney Monk, Terrell McCalister, Teddy Heckman, Gator Thompson, Dave Blatz

Historical Consultants: Big Thicket National Preserve & Visitors Center

Video Production: Colmesneil ISD Tricaster Team - Advisor Kathy Gobert

PROPERTY DIRECTOR
John Cooley

Committee Members:
Josh McClure, Stephen Hill, Michael Evans, Phillip Scoggins, Joseph Conner

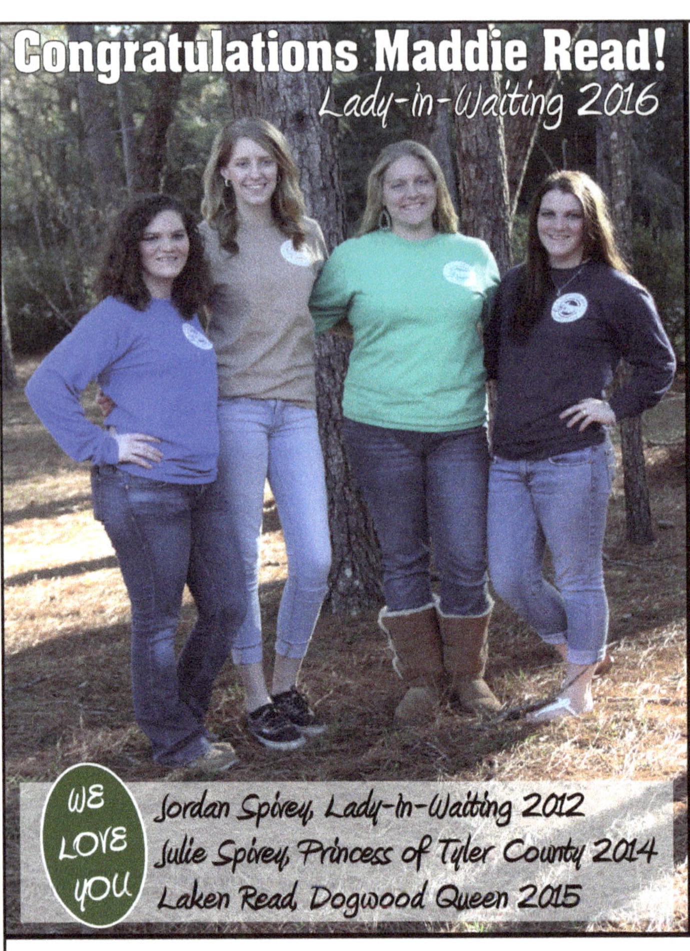

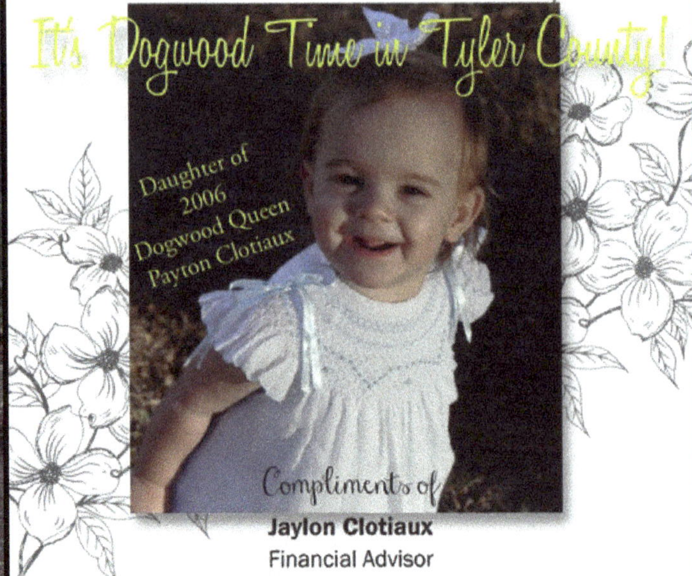

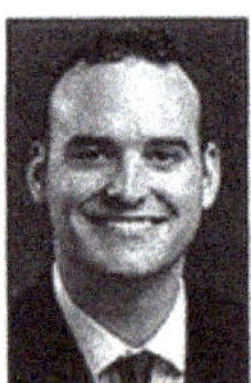

Jaylon Clotiaux
Financial Advisor

Edward Jones
MAKING SENSE OF INVESTING

805 South Wheeler
Jasper, TX 75951
Bus. 409-384-7005 TF. 800-441-7869
Fax 888-639-5914
jaylon.clotiaux@edwardjones.com
www.edwardjones.com

LITTLE SPROUTS CHILDCARE

508 West Elder • Colmesneil
409-837-5104

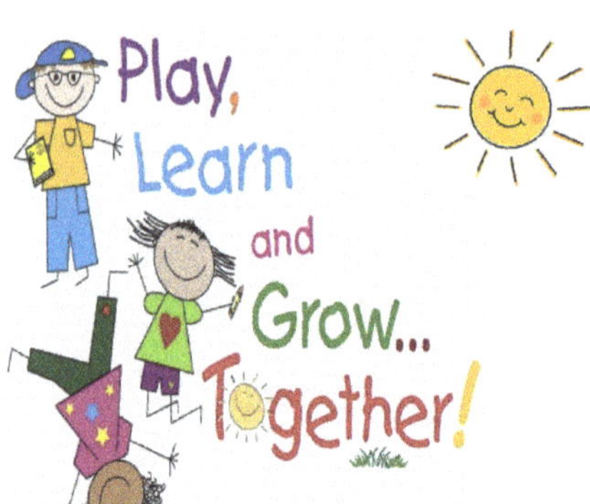

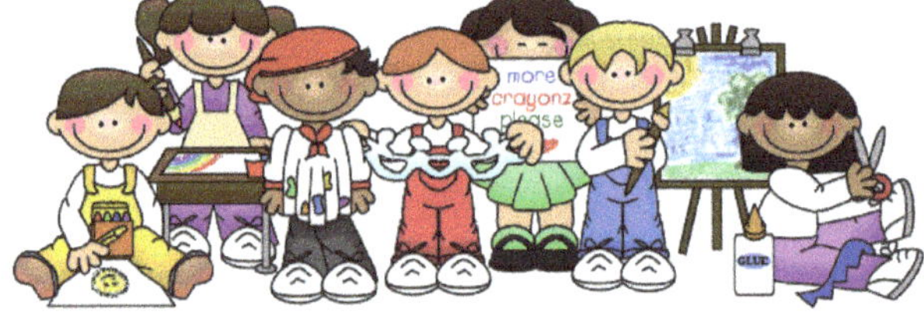

At Little Sprouts we believe that quality child care facilitates the healthy growth and development of the "whole" child. We believe that all children are different and unique; our environment accepts and encourages children to express their full potential in all aspects of their development.

Our program offers each child a balance of challenging, stimulating experiences, and experiences that are warm and secure. It is essential to achieve the maximum benefit from an experience, a child must experience that they are loved, valued, and cared for.

Graham's Grill & Grocery

Colmesneil

409-837-2014 • Call-ins Welcome
OPEN 7 DAYS A WEEK 5AM-8PM

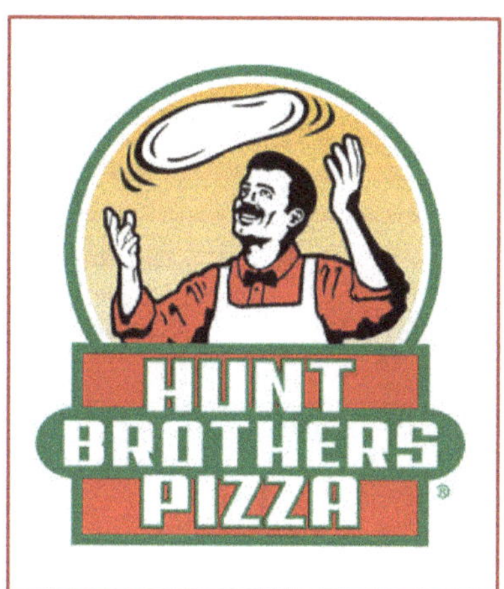

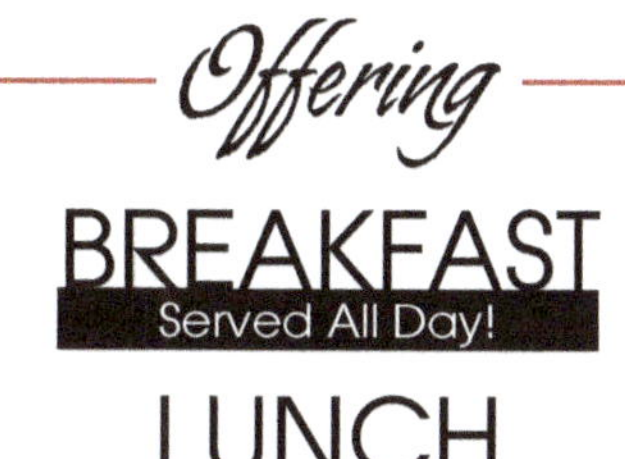

Offering

BREAKFAST
Served All Day!

LUNCH

DINNER

Steam Table
Homestyle Cooking

- **FISHING & HUNTING LICENSES**
- **Tackle**
- **Feed**

Grocery Items • Fresh Vegetables • Deli Items

Karolyna Kuts
SALON   SPA & BOUTIQUE

1215 West Bluff
WOODVILLE - 409-331-9950
Open: Mon by Appt - Tues-Fri 9-5- Sat 9-2

Karolyna Kuts is a unique boutique with gypsy souled gals in a fun loving atmosphere that is more than your average salon, with vendors that offer a wonderful shopping experience.

We have a variety of collectibles, gifts, fashion accessories, home décor, 'one of a kind' pieces of jewelry and a whole lot more.

HAIRSTYLISTS:

Loretta

Meghan

LeeAnna

NAIL TECH:

Jessica

Formerly Loretta's

COME SEE
Loretta
on Thursday's

We LOVE to make you feel your **BEST!**

Salon Services:
Haircuts
Shampoo / Style
Blow Dry / Flat Iron
Perms
Highlights / Lowlights
All Color Needs
Smoothing Treatments
Deep Conditioning
   Treatments
All Hair Types
Extensions

Packages Available for Special Occasion

Spa Services:

NAILS
Manicures
Pedicures
Acrylics
Etc

SKIN CARE
Basic Facial
Basic-Full Face Makeup
Eyelash Extensions

WAXING
Full Face
Arms
Legs

Call For Additional Services Not Listed!
LIKE US ON FACEBOOK

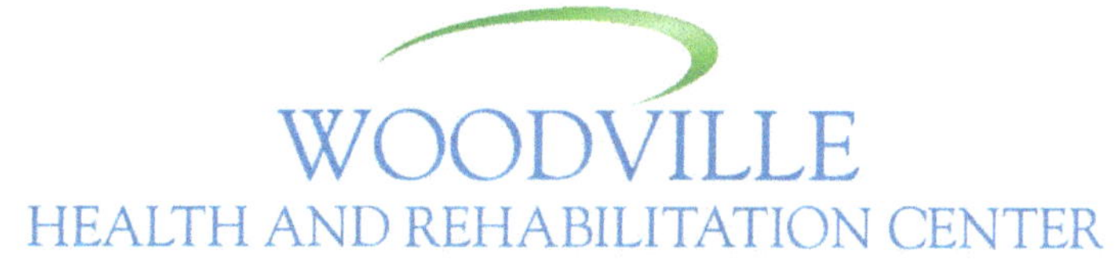

Cantex Continuing Care Network has been performing the highest standards of transitional healthcare and long-term residential services for more than 30 years.

We are committed to excellence!

- Private rooms available
- Wi-Fi internet access
- Flat screen television

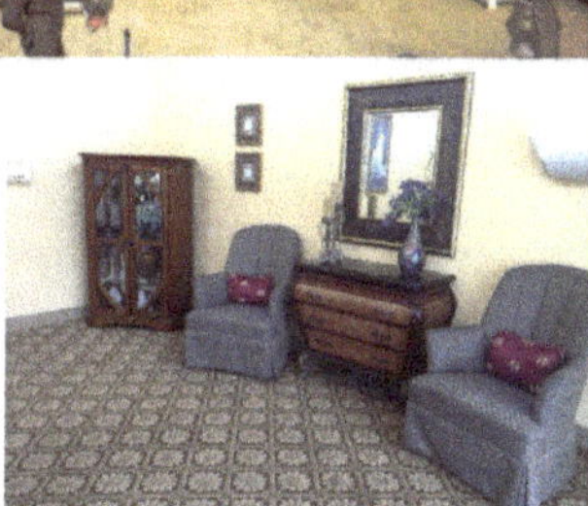
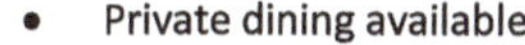

- Private dining available
- Restaurant style dining
- Salon services
- Housekeeping
- Transportation

Medicare & Medicaid Certified
Most insurances accepted
Private Pay

102 North Beech Street, Woodville, TX 75979
Phone: 409-283-2555   Fax: 409-283-8446
www.cantexcc.com

Rehabilitation
Hospice Care
Long Term Care
Respite Care

Physical Therapy
Occupational Therapy
Speech Therapy
Respiratory Therapy

Post-Operative Care
Care Transitions
Intravenous Therapy
Wound Care
Oxygen Therapy
Stroke Recovery
Medication Administration
Tracheostomy Care
Feeding tubes
Pain Management
COPD
Bladder Control
Lymphedema Management
Contractures
Neuromuscular Conditions
Balance Recovery
Swallow Function Recovery
Inflammatory Conditions
Rheumatoid Arthritis

....and more!

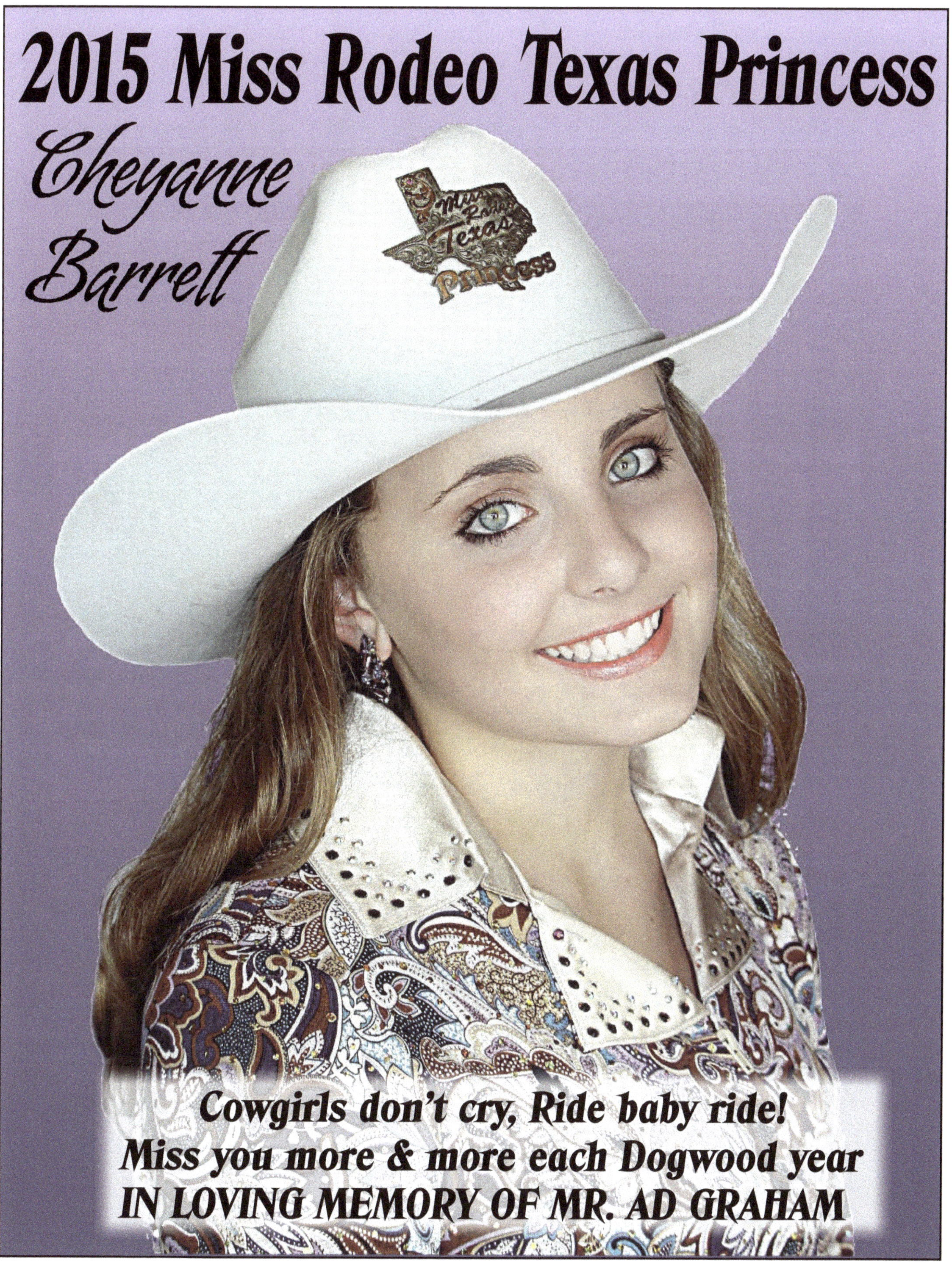

2015 Miss Rodeo Texas Princess
Cheyanne Barrett
Cowgirls don't cry, Ride baby ride!
Miss you more & more each Dogwood year
IN LOVING MEMORY OF MR. AD GRAHAM

Compliments of

ROTARY CLUB
of Woodville

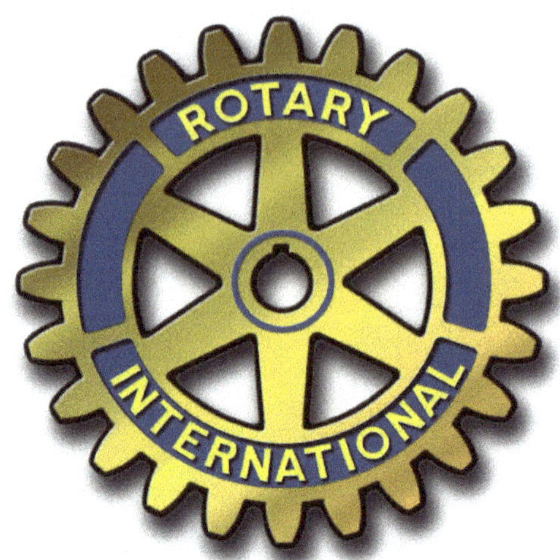

ROTARY
INTERNATIONAL

Mankind is our business

"Service Above Self"

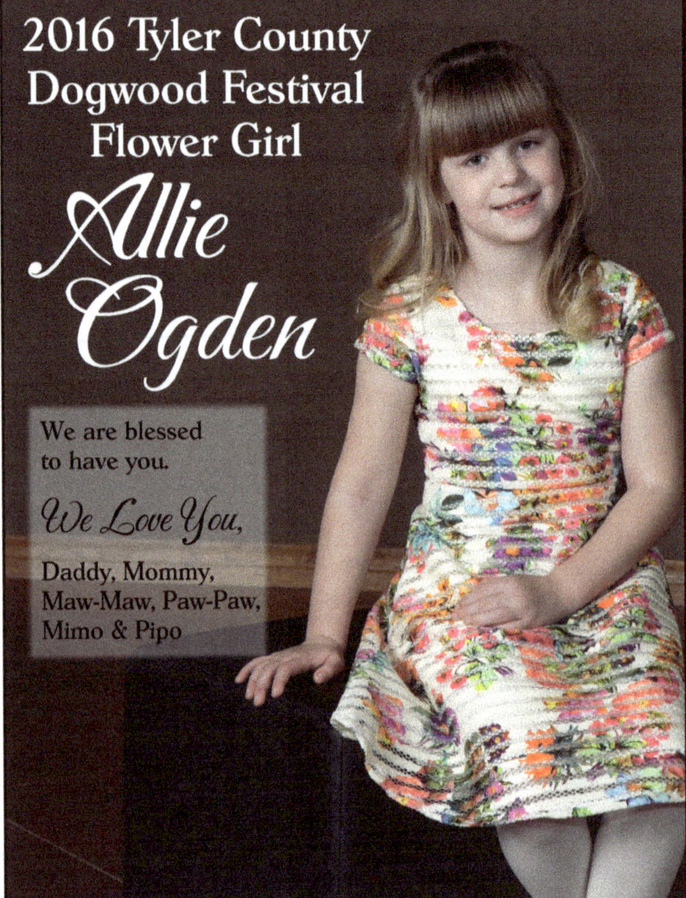

2016 Tyler County
Dogwood Festival
Flower Girl

Allie
Ogden

We are blessed
to have you.

We Love You,

Daddy, Mommy,
Maw-Maw, Paw-Paw,
Mimo & Pipo

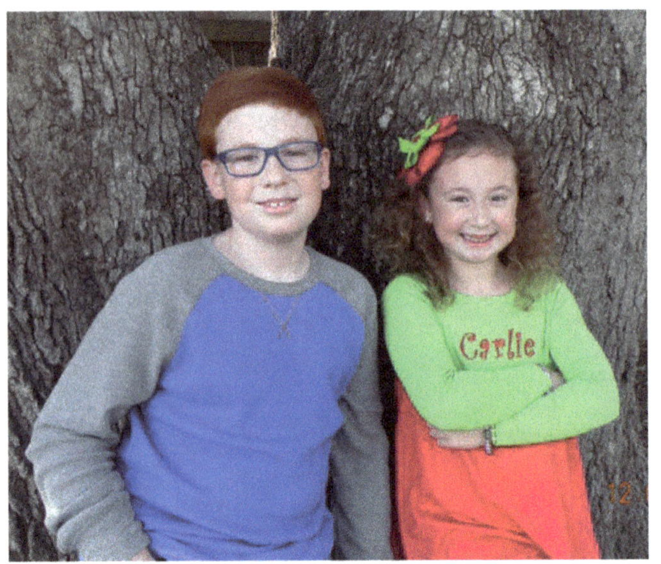

Carlie

Compliments of

J. Michael
Risinger

Attorney at Law

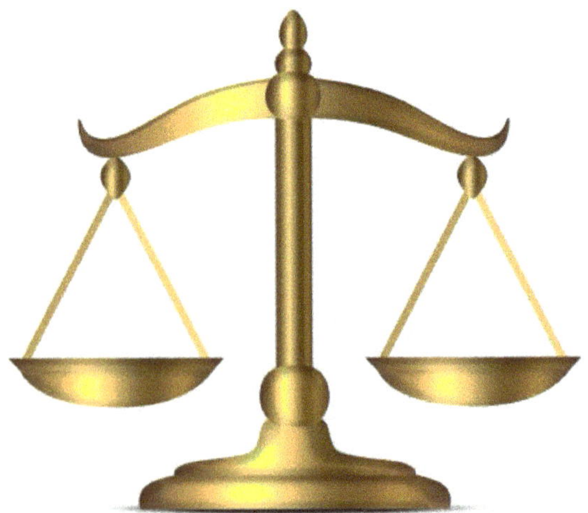

314 South Magnolia
WOODVILLE

409-283-2324